STATISTICS FROM A TO Z

STATISTICS FROM A TO Z
Confusing Concepts Clarified

ANDREW A. JAWLIK

Published by John Wiley & Sons, Inc., Hoboken, New Jersey.
Published simultaneously in Canada.

For general information on our other products and services or for technical support, please contact our Customer Care Department within the United States at (800) 762-2974, outside the United States at (317) 572-3993 or fax (317) 572-4002.

Wiley also publishes its books in a variety of electronic formats. Some content that appears in print may not be available in electronic formats. For more information about Wiley products, visit our web site at www.wiley.com.

Library of Congress Cataloging-in-Publication Data

Names: Jawlik, Andrew.
Title: Statistics from A to Z : confusing concepts clarified / Andrew Jawlik.
Description: Hoboken, New Jersey : John Wiley & Sons, Inc., [2016].
Identifiers: LCCN 2016017318 | ISBN 9781119272038 (pbk.) | ISBN 9781119272007 (epub)
Subjects: LCSH: Mathematical statistics–Dictionaries. | Statistics–Dictionaries.
Classification: LCC QA276.14 .J39 2016 | DDC 519.503–dc23
LC record available at https://lccn.loc.gov/2016017318

10 9 8 7 6 5 4 3 2 1

To my wonderful wife, Jane, who is a 7 Sigma*.

<hr>

*See the article, "Sigma", in this book.

CONTENTS

OTHER CONCEPTS COVERED IN THE ARTICLES

1-Sided or 1-Tailed: see the articles *Alternative Hypothesis* and *Alpha, α*.

1-Way: an analysis that has one Independent (x) Variable, e.g., 1-way ANOVA.

2-Sided or 2-Tailed: see the articles *Alternative Hypothesis* and *Alpha, α*.

2-Way: an analysis that has two Independent (x) Variables, e.g., 2-way ANOVA.

68-95-99.7 Rule: same as the Empirical Rule. See the article *Normal Distribution*.

Acceptance Region: see the article *Alpha, α*.

Adjusted R^2: see the article *r, Multiple R, r^2, R^2, R Square, R^2 Adjusted*.

aka: also known as.

Alias: see the article *Design of Experiments (DOE) – Part 2*.

Associated, Association: see the article *Chi-Square Test for Independence*.

Assumptions: requirements for being able to use a particular test or analysis. For example, ANOM and ANOVA require approximately Normal data.

Attributes data, Attributes Variable: same as Categorical or Nominal data or Variable. See the articles *Variables* and *Chi-Square Test for Independence*.

Autocorrelation: see the article *Residuals*.

Average Absolute Deviation: see the article *Variance*.

Average: same as the Mean – the sum of a set of numerical values divided by the Count of values in the set.

Bernoulli Trial: see the article *Binomial Distribution.*

Beta: the probability of a Beta Error. See the article *Alpha and Beta Errors.*

Beta Error: featured in the article *Alpha and Beta Errors.*

Bias: see the article *Sample, Sampling.*

Bin, Binning: see the articles *Chi-Square Test for Goodness of Fit* and *Charts/Graphs/Plots – Which to Use When.*

Block, Blocking: see the article *Design of Experiments (DOE) – Part 3.*

Box Plot, Box and Whiskers Plot: see the article *Charts/Graphs/Plots – Which to Use When.*

C_m, C_p, C_r, or C_{PK}: see the article *Process Capability Analysis (PCA).*

Capability, Capability Index: see the article *Process Capability Analysis (PCA).*

Categorical data, Categorical Variable: same as Attribute or Nominal data/Variable. See the articles *Variables* and *Chi-Square Test for Independence.*

CDF: see Cumulative Density Function.

Central Limit Theorem: see the article *Normal Distribution.*

Central Location: same as Central Tendency. See the article *Distributions – Part 1: What They Are.*

Central Tendency: same as Central Location. See the article *Distributions – Part 1: What They Are.*

Chebyshev's Theorem: see the article *Standard Deviation.*

Confidence Coefficient: same as Confidence Level. See the article *Alpha, α.*

Confidence Level: (aka Level of Confidence aka Confidence Coefficient) equals 1 – Alpha. See the article *Alpha, α.*

Confounding: see the article *Design of Experiments (DOE) – Part 3.*

Contingency Table: see the article *Chi-Square Test for Independence.*

Continuous data or Variables: see the articles *Variables* and *Distributions – Part 3: Which to Use When.*

Control, "in … " or "out of … ": see the article *Control Charts – Part 1: General Concepts and Principles.*

Control Limits, Upper and Lower: see the article *Control Charts – Part 1: General Concepts and Principles.*

Count data, Count Variables: aka Discrete data or Discrete Variables. See the article *Variables*.

Covariance: see the article *Correlation – Part 1*.

Criterion Variable: see the article *Variables*.

Critical Region: same as Rejection Region. See the article *Alpha, α*.

Cumulative Density Function (CDF): the formula for calculating the Cumulative Probability of a Range of values of a Continuous random Variable, for example, the Cumulative Probability that $x \leq 0.5$.

Cumulative Probability: see the article *Distributions – Part 2: How They Are Used*.

Curve Fitting: see the article *Regression – Part 5: Simple Nonlinear*.

Dependent Variable: see the article *Variables*.

Descriptive Statistics: see the article *Inferential Statistics*.

Dot Plot: see the article *Charts/Graphs/Plots – Which to Use When*.

Deviation: the difference between a data value and a specified value (usually the Mean). See the article *Regression – Part 1: Sums of Squares*. See also the article *Standard Deviation*.

Discrete data or Variables: see the articles *Variables* and *Distributions – Part 3: Which to Use When*.

Dispersion: see the article *Variation/Variability/Dispersion/Spread* (they all mean the same thing).

Effect Size: see the article *Power*.

Empirical Rule: same as the 68-95-99.7 Rule. See the article *Normal Distribution*.

Expected Frequency: see the articles *Chi-Square Test for Goodness of Fit* and *Chi-Square Test for Independence*.

Expected Value: see the articles *Chi-Square Test for Goodness of Fit* and *Chi-Square Test for Independence*.

Exponential: see the article *Exponential Distribution*.

Exponential Curve: see the article *Regression – Part 5: Simple Nonlinear*.

Exponential Transformation: see the article *Regression – Part 5: Simple Nonlinear*.

Extremes: see the article *Variation/Variability/Dispersion/Spread*.

F-test: see the article *F*.

Factor: see the articles *ANOVA – Parts 3 and 4* and *Design of Experiments (DOE) – Part 1*.

False Positive: an Alpha or Type I Error; featured in the article *Alpha and Beta Errors*.

False Negative: a Beta or Type II Error; featured in the article *Alpha and Beta Errors*.

Frequency: a Count-like Statistic which can be non-integer. See the articles *Chi-Square Test for Goodness of Fit* and *Chi-Square Test for Independence*.

Friedman Test: see the article *Nonparametric*.

Gaussian Distribution: same as Normal Distribution.

Generator: see the article *Design of Experiments (DOE) – Part 3*.

Goodness of Fit: see the articles *Regression – Part 1: Sums of Squares* and *Chi-Square Test for Goodness of Fit*.

Histogram: see the article *Charts/Graphs/Plots – Which to Use When*.

Independence: see the article *Chi-Square Test for Independence*.

Independent Variable: see the article *Variables*.

Interaction: see the articles *ANOM*; *ANOVA – Part 4: 2-Way*; *Design of Experiments, Parts 1, 2, and 3*; *Regression – Part 4: Multiple Linear*.

Intercept: see the article *Regression – Part 2: Simple Linear*.

InterQuartile Range (IQR): see the article *Variation/Variability/Dispersion/Spread*.

Kruskal–Wallis Test: see the article *Nonparametric*.

Kurtosis: a measure of the Shape of a Distribution. See the article *Distributions – Part 1: What They Are*.

Least Squares: (same as Least Sum of Squares or Ordinary Least Sum of Squares) see the articles *Regression – Part 1: Sums of Squares* and *Regression – Part 2: Simple Linear*.

Least Sum of Squares: same as Least Squares.

Level of Confidence: same as Confidence Level; equal to $1 - \alpha$. See the article *Alpha, α*.

Level of Significance: same as Significance Level, Alpha (α). See the articles *Alpha, α* and *Statistically Significant*.

Line Chart: see the article *Charts/Graphs/Plots – Which to Use When*.

Logarithmic Curve, Logarithmic Transformation: see the article *Regression – Part 5: Simple Nonlinear*.

Main Effect: a Factor which is not an Interaction. See the articles *ANOVA – Part 4: 2-Way* and *Design of Experiments (DOE) – Part 2*.

Mann–Whitney Test: see the article *Nonparametric*.

Mean: the average. Along with Mean and Median, it is a measure of Central Tendency.

Mean Absolute Deviation (MAD): see the article *Variation/Variability/Dispersion/Spread*.

Mean Sum of Squares: see the article *ANOVA – Part 2* (MSB and MSW) and the article *F*.

Measurement data: same as Continuous data.

Median: the middle of a range of values. Along with Mean and Mode, it is a measure of Central Tendency. It is used instead of the Mean in Nonparametric Analysis. See the article *Nonparametric*.

Memorylessness: see the article *Exponential Distribution*.

Mode: the most common value within a group (e.g., a Sample or Population, or Process). There can be more than one Mode. Along with Mean and Median, Mode is a measure of Central Tendency.

MOE: see the article *Margin of Error*.

MSB and MSW: see the article *ANOVA – Part 2* (MSB and MSW) and the article *F*.

Multiple R: see the article *r, Multiple R, r^2, R^2, R Square, R^2 Adjusted*.

Multiplicative Law of Probability: see the article *Chi-Square Test for Independence*.

Nominal data, Nominal Variable: same as Categorical or Attributes data or Variable. See the article *Variables*.

One-Sided, One-Tailed: (same as 1-sided, 1-tailed) see the articles *Alternative Hypothesis* and *Alpha, α*.

One-Way: same as 1-Way; an analysis that has one Independent (x) Variable. For example, 1-way ANOVA.

Outlier: see the article *Variation/Variability/Dispersion/Spread*.

Parameter: a measure of a property of a Population or Process, e.g., the Mean or Standard Deviation. The counterpart for a Sample is called a "Statistic." Parameters are usually denoted by characters in the Greek Alphabet, such as μ or σ.

Parametric: see the article *Nonparametric*.

Pareto Chart: see the article *Charts/Graphs/Plots – Which to Use When*.

PCA: see the article *Process Capability Analysis (PCA)*.

PDF: see Probability Density Function.

Pearson's Coefficient, Pearson's *r*: the correlation Coefficient, *r*. See the article *Correlation – Part 2*.

Performance Index: see the article *Process Capability Analysis (PCA)*.

PMF: see Probability Mass Function.

Polynomial Curve: see the article *Regression – Part 5: Simple Nonlinear*.

"Population or Process": where most texts say "Population," this book adds "or Process." Ongoing Processes are handled the same as Populations, because new data values continue to be created. Thus, like Populations, we don't have complete data for ongoing Processes.

Power Transformation: see the article *Regression – Part 5: Simple Nonlinear*.

Probability Density Function (PDF): the formula for calculating the Probability of a single value of a Continuous random Variable of, for example, the Probability that $x = 5$. (For Discrete random Variables, the corresponding term is Probability Mass Function, PMF.) See also Cumulative Density Function.

Probability Distribution: see the article *Distributions – Part 1: What They Are*.

Probability Mass Function (PMF): the formula for calculating the Probability of a single value of a Discrete random Variable of, for example, the Probability that $x = 5$.

Qualitative Variable/Qualitative data: same as Categorical Variable and Categorical data. See the articles *Variables* and *Chi-Square Test for Independence*.

Outlier: see the article *Variation/Variability/Dispersion/Spread*.

Random Sample: see the article *Sample, Sampling*.

Random Variable: see the article *Variables*.

Range: see the article *Variation/Variability/Dispersion/Spread*.

Rational Subgroup: see the article *Control Charts – Part 1*.

Rejection Region: same as Critical Region. See the article *Alpha, α*.

Replacement, Sampling With or Without: see the article *Binomial Distribution*.

Resolution: see the article *Design of Experiments (DOE) – Part 3*.

Response Variable: see the articles *Variables* and *Design of Experiments (DOE) – Part 2*.

Run Rules: see the article *Control Charts – Part 1*.

Scatterplot: see the article *Charts/Graphs/Plots – Which to Use When*.

Shape: see the article *Distributions – Part 1: What They Are*.

Significance Level: see the article *Alpha, α*.

Significant: see the article *Statistically Significant*.

Slope: see the article *Regression – Part 2: Simple Linear*.

Spread: see the article *Variation/Variability/Dispersion/Spread*.

Standard Normal Distribution: see the articles *Normal Distribution* and *z*.

Statistic: a measure of a property of a Sample, e.g., the Mean or Standard Deviation. The counterpart for a Population or Process is called a "Parameter." Statistics are usually denoted by characters based on the Roman Alphabet, such as $\bar{x}$ or *s*.

Statistical Inference: same as Inferential Statistics; see the article by that name.

Statistical Process Control: see the article *Control Charts – Part 1: General Concepts and Principles*.

Student's *t*: see the article *t, The Test Statistic and Its Distributions*.

Tail: see the articles *Alpha, α* and *Alternative Hypothesis*.

Three Sigma Rule: same as Empirical Rule and the 68-95-99.7 Rule. See the article *Normal Distribution*.

Transformation: see the article *Regression – Part 5: Simple Nonlinear*.

Two-Sided, Two-Tailed: same as 2-Sided, 2-Tailed. See the articles *Alpha, α* and *Alternative Hypothesis*.

Two-way: same as 2-Way; an analysis that has two Independent (*x*) Variables, e.g., 2-way ANOVA.

Type I and Type II Errors: same as Alpha and Beta Errors, respectively. See the article by that name.

Variables data: same as Continuous data. See the articles *Variables* and *Distributions – Part 3: Which to Use When*.

Variability: see the article *Variation/Variability/Dispersion/Spread*.

Wilcoxon Test: see the article *Nonparametric*.

WHY THIS BOOK IS NEEDED

A statistician responds to a marriage proposal.

Statistics can be confusing – even for smart people, and even for smart technical people.

As an illustration, how quickly can we figure out whether the woman pictured above agreed to get married? (For the answer, see the article in this book, *"Fail to Reject the Null Hypothesis."*)

This is understandable, not only because **some of the concepts are inherently complicated and difficult to understand**, but also because:

- Different terms are used to mean the same thing

For example, the Dependent Variable, the Outcome, the Effect, the Response, and the Criterion are all the same thing. And – believe it or not – there are at least seven different names and 18 different acronyms used for just the three Statistics: Sum of Squares Between, Sum of Squares Within, and Sum of Squares Total.

Synonyms may be wonderful for poets and fiction writers, but they confuse things unnecessarily for students and practitioners of a technical discipline.

- Conversely, a single term can have very different meanings

For example, "SST" is variously used for "Sum of Squares Total" or "Sum of Squares Treatment." (The latter is actually a component part of the former.)

- Sometimes, there is no single "truth"

The acknowledged experts sometimes disagree on fundamental concepts. For example, some experts specify the use of the Alternative Hypothesis in their methods of Hypothesis Testing. Others are "violently opposed" to its use. Other experts recommend avoiding Hypothesis Testing completely, because of the confusing language.

- Words can have different meanings from their usage in everyday language

The meaning of words in statistics can sometimes be very different from, or even the opposite of, the meaning of the same words in normal, everyday language.

For example, in a Bernoulli experiment on process quality, a quality failure is called a "success." Also, for Skew or Skewness, in statistics, "left" means right.

Everyday language:
"Skewed to the left."

Statistics:
Skewed to the right,
positive skew

- A confusing array of choices

Which Distribution do I use when? Which Test Statistic? Which test? Which Control Chart? Which type of graph?

There are several choices for each – some of which are good in a given situation, some not.

- And the existing books don't seem to make things clear enough

Even those with titles targeting the supposedly clueless reader do not provide sufficient explanation to clear up a lot of this confusion. Students and professionals continue to look for a book which would give them a true intuitive understanding of statistical concepts.

Also, if you look up a concept in the index of other books, you will find something like this:

"Degrees of freedom, 60, 75, 86, 91–93, 210, 241"

So, you have to go to six different places, pick up the bits and pieces from each, and try to assemble for yourself some type of coherent concept. In this book, each concept is completely covered in one or more contiguous short articles (usually three to seven pages each). And we don't need an index, because you find the concepts alphabetically – as in a dictionary or encyclopedia.

WHAT MAKES THIS BOOK UNIQUE?

It is much **easier to understand** than other books on the subject, because of the following:

- **Alphabetically arranged**, like a mini-encyclopedia, for immediate access to the specific knowledge you need at the time.
- **Individual articles** which completely treat one concept per article (or series of contiguous articles). No paging through the book for bits and pieces here and there.

    Almost all the articles start with a one-page summary of five or so **Keys to Understanding**, which gives you **the whole picture on a single page**. The remaining pages in the article provide a more in-depth explanation of each of the individual keys.
- **Unique graphics that teach:**
    - **Concept Flow Diagrams:** visually depict how one concept leads to another and then another in the step-by-step thought process leading to understanding.
    - **Compare-and-Contrast Tables:** for reinforcing understanding via differences, similarities, and any interrelationships between related concepts – e.g., p vs. Alpha, z vs. t, ANOVA vs. Regression, Standard Deviation vs. Standard Error.
    - **Cartoons** to enhance "rememberability."

- **Highest ratio of visuals to text** – plenty of pictures and diagrams and tables. This provides more concrete reinforcement of understanding than words alone.
- **Visual enhancing of text** to increase focus and to improve "rememberability." All statistical terms are capitalized. Extensive use of short paragraphs, numbered items, bullets, bordered text boxes, arrows, underlines, and bold font.
- **Repetition:** An individual concept is often explained in several ways, coming at it from different aspects. If an article needs to refer to some content covered in a different article, that content is usually repeated within the first article, if it's not too lengthy.
- A **Which Statistical Tool to Use** article: Given a type of problem or question, which test, tool, or analysis to use. In addition, there are individual **Which to Use When** articles for Distributions, Control Charts, and Charts/Graphs/Plots.

Wider Scope – Statistics I and Statistics II and Six Sigma Black Belt. Most books are focused on statistics in the social sciences, and – to a lesser extent – physical sciences or management. They don't cover statistical concepts important in process and quality improvement (Six Sigma or industrial engineering).

Authored by a recent student, who is freshly aware of the statistical concepts that confused him – and why. (The author recently completed a course of study for professional certification as a Lean Six Sigma black belt – a process and quality improvement discipline which uses statistics extensively. He had, years earlier, earned an MS in Mathematics in a concentration which did not include much statistics content.)

HOW TO USE THIS BOOK

Use this book when:

- you're confused about a specific statistical concept or which statistical tool to use
- you need a refresher on a statistical concept or method, just to be sure
- you want help in making things easier to understand when communicating with others

It can be useful:

- while studying or while taking an open-book exam
- on the job
- as a reference, when developing presentations or writing e-mails

To find a subject, you can flip through the book like an old dictionary or encyclopedia volume. If the subject you are looking for does not have an article devoted to it, there is likely a glossary description for it. And/or it may be covered in an article on another subject. In an alphabetically-organized book like this, the Contents and the Other Concepts pages make an Index unnecessary.

See the Contents at the beginning of this book for a list of the articles covering the major concepts. Following the Contents is a section called "Other Concepts Covered in the Articles." Here, you can find concepts which do not headline their own articles, for example:

Acceptance Region: see the article *Alpha, α*.

If you have a statistical problem to solve or question to answer and don't know how to go about it, see the article **Which Statistical Tool to Use to Solve Some Common Problems**. There are also **Which to Use When** articles for Distributions, Control Charts, and Charts/Graphs/Plots.

This book is designed for use as a reference for looking up specific topics, not as a textbook to be read front-to-back. However, if you do want to use this book as a single source for learning statistics, not just a reference, you could read the following articles in the order shown:

- *Inferential Statistics*
- *Alpha, p, Critical Value, and Test Statistic – How They Work Together*
- *Hypothesis Testing, Parts 1 and 2*
- *Confidence Intervals, Parts 1 and 2*
- *Distributions, Parts 1 – 3*
- *Which Statistical Tool to Use to Solve Some Common Problems*
- Articles on individual tests and analyses, such as *t-Tests*, *F*, *ANOVA*, and *Regression*

At the end of these and all other articles in the book is a list of **Related Articles** which you can read for more detail on related subjects.

ALPHA, α

Summary of Keys to Understanding

 1. In Inferential Statistics, p is the Probability of an Alpha ("False Positive") **Error.**

 2. Alpha is the highest value of p that we are willing to tolerate and still say that a difference, change, or effect observed in the Sample is "Statistically Significant."

I want to be 95% confident of avoiding an Alpha Error. So, I'll select $\alpha = 5\%$.

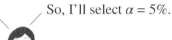

 3. Alpha is a Cumulative Probability, represented as an area under the curve, at one or both tails of a Probability Distribution. p is also a Cumulative Probability.

Areas under the curve (right tail)
α:☐ p:▨

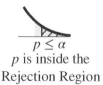 

Acceptance Region

Rejection Region (α) $p \leq \alpha$ $p > \alpha$
   p is inside the p extends into the
   Rejection Region   Acceptance Region

 4. In Hypothesis Testing, if $p \leq \alpha$, Reject the Null Hypothesis. If $p > \alpha$, Accept (Fail to Reject) the Null Hypothesis.

 5. Alpha defines the Critical Value(s) of Test Statistics, such as z, t, F, or Chi-Square. The Critical Value or Values, in turn, define the Confidence Interval.

Statistics from A to Z: Confusing Concepts Clarified, First Edition. Andrew A. Jawlik.
© 2016 John Wiley & Sons, Inc. Published 2016 by John Wiley & Sons, Inc.

Explanation

> **1. In Inferential Statistics, p is the Probability of an Alpha** ("False Positive") **Error.**

In Inferential Statistics, we use data from a Sample to estimate a property (say, the Mean) of the Population or Process from which the Sample was taken. **Being an estimate, there is a risk of error.**

One type of error is the Alpha Error (also known as "Type I Error" or "False Positive").

I saw a unicorn.

Alpha Error
(False Positive)

An Alpha Error is the error of seeing something which is not there, that is, **concluding that there is** a Statistically Significant difference, change, or effect, **when in fact there is not.** For example,

- Erroneously concluding that there is a difference in the Means of two Populations, when there is not, or
- Erroneously concluding that there has been a change in the Standard Deviation of a Process, when there has not, or
- Erroneously concluding that a medical treatment has an effect, when it does not.

In Hypothesis Testing, the Null Hypothesis states that there is no difference, change, or effect. All these are examples of **Rejecting the Null Hypothesis when the Null Hypothesis is true.**

p is the Probability of an Alpha Error, a "False Positive."

It is calculated as part of the Inferential Statistical analysis, for example, in a t-test or ANOVA.

How does an Alpha Error happen? **An Alpha Error occurs when data in our Sample are <u>not</u> representative of the overall Population or Process from which the Sample was taken.**

If the Sample Size is large enough, the great majority of Samples of that size will do a good job of representing the Population or Process. However, some won't. p **tells us how probable it is that our Sample is unrepresentative enough to produce an Alpha Error.**

2. Alpha is the highest value of p that we are willing to tolerate and still say that a difference, change, or effect observed in the Sample is "Statistically Significant."

In this article, we use Alpha both as an adjective and as a noun. This might cause some confusion, so let's explain.

"Alpha," as an adjective, describes a type of error, the Alpha Error. Alpha as a noun is something related, but different.

First of all, what it is not: **Alpha**, as a noun, **is not**

- a Statistic or a Parameter, which describes a property (e.g., the Mean) of a Sample or Population
- a Constant, like those shown in some statistical tables.

Second, what it is: Alpha, as a noun, **is**

- **a value of p which defines the boundary of the values of p which we are willing to tolerate from those which we are not.**

For example, if we are willing to tolerate a 5% risk of a False Positive, then we would select $\alpha = 5\%$. That would mean that we are willing to tolerate $p \leq 5\%$, but not $p > 5\%$.

Alpha must be selected prior to collecting the Sample data. This is to help ensure the integrity of the test or experiment. If we have a look at the data first, that might influence our selection of a value for Alpha.

Rather than starting with Alpha, it's probably more natural to think in terms of a Level of Confidence first. Then we subtract it from 1 (100%) to get Alpha.

If we want to be 95% sure, then we want a 95% Level of Confidence (aka "Confidence Level").

By definition, $\alpha = \mathbf{100\% - Confidence\ Level.}$ (And, so Confidence Level $= 100\% - \alpha$.)

I want to be 95% confident of avoiding an Alpha Error.   So, I'll select $\alpha = 5\%$.

Alpha is called the "Level or Significance" or "Significance Level."

- **If p is calculated to be less than or equal to the Significance Level, α, then any observed difference, change, or effect calculated from our Sample data is said to be "Statistically Significant."**

- **If $p > \alpha$, then it is <u>not</u> Statistically Significant.**

Popular choices for Alpha are 10% (0.1), 5% (0.05), 1% (0.01), 0.5% (0.005), and 0.1% (0.001). But, why wouldn't we always select as low a level of Alpha as possible? Because, **the choice of Alpha is a tradeoff between Alpha (Type I) Error and Beta (Type 2) Error** – or put another way – between a False Positive and a False Negative. If you reduce the chance (Probability) of one, you increase the chance of the other.

Choosing $\alpha = 0.05$ (5%) is generally accepted as a good balance for most uses. The pros and cons of various choices for Alpha (and Beta) in different situations are covered in the article, *Alpha and Beta Errors*.

> **3. Alpha is a Cumulative Probability, represented by an area under the curve, at one or both tails of a Probability Distribution. p is also a Cumulative Probability.**

Below are diagrams of the Standard Normal Distribution. The Variable on its horizontal axis is the Test Statistic, z. Any point on the curve is the Probability of the value of z directly below that point.

Probabilities of individual points are usually less useful in statistics than Probabilities of ranges of values. The latter are called Cumulative Probabilities. **The Cumulative Probability of a range of values is calculated as the area under the curve above that range of values.** The Cumulative Probability of all values under the curve is 100%.

We start by selecting a value for Alpha, most commonly 5%, which tells us how big the shaded area under the curve will be. **Depending on the type of problem we're trying to solve, we position the shaded area (α) under the left tail, the right tail, or both tails.**

left-tailed: our orders ship in less than 4 days   2-tailed: our estimate of the Mean is accurate   right-tailed: our light bulbs last longer than 1300 hours

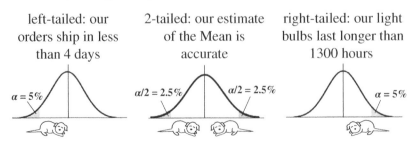

If it's one tail only, the analysis is called "1-tailed" or "1-sided" (or "left-tailed or "right-tailed"), and Alpha is entirely under one side of the curve. If it's both tails, it's called a "2-tailed" or **"2-sided" analysis. In that case, we divide Alpha by two, and put half under each tail.** For more on tails, see the article *Alternative Hypothesis*.

There are two main methods in Inferential Statistics – Hypothesis Testing and Confidence Intervals. Alpha plays a key role in both. First, let's take a look at Hypothesis Testing:

 | **4. In Hypothesis Testing, if $p \leq \alpha$, Reject the Null Hypothesis. If $p > \alpha$, Accept (Fail to Reject) the Null Hypothesis.**

In Hypothesis testing, p is compared to Alpha, in order to determine what we can conclude from the test.

Hypothesis Testing starts with a **Null Hypothesis** – a statement that there is no (Statistically Significant) difference, change, or effect.

We select a value for Alpha (say 5%) and then collect a Sample of data. Next, a statistical test (like a *t*-test or *F*-test) is performed. The test output includes a value for p.

p **is** the Probability of an Alpha Error, a False Positive, that is, **the Probability that any difference, effect, or change shown by the Sample data is not Statistically Significant.**

If p is small enough, then we can be confident that there really is a difference, change, or effect. How small is small enough? **Less than or equal to Alpha.** Remember, we picked Alpha as the upper boundary for the values of p which indicate a tolerable Probability of an Alpha Error. So, $p > \alpha$ is an unacceptably high Probability of an Alpha Error.

How confident can we be? As confident as the Level of Confidence. For example, with a 5% Alpha (Significance Level), we have a 100% – 5% = 95% Confidence Level. So, ...

If $p \leq \alpha$, then we conclude that:

- the Probability of an Alpha Error is within the range we said we would tolerate, so **the observed difference, change, or effect we are testing is Statistically Significant.**
- **in a Hypothesis test, we would Reject the Null Hypothesis.**
- **the smaller the p-value, the stronger the evidence for this conclusion.**

How does this look graphically? Below are three close-ups of the right tail of a Distribution. This is for a 1-tailed test, in which the shaded area represents Alpha and the hatched areas represent p. (In a 2-tailed test, the left and right tails would each have $\alpha/2$ as the shaded areas.)

- Left graph below: in Hypothesis Testing, some use the term "Acceptance Region" or "Non-critical Region" for the unshaded white area under the Distribution curve, and "Rejection Region" or "Critical Region" for the shaded area representing Alpha.

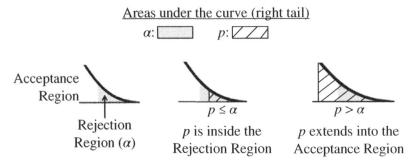

Areas under the curve (right tail)

α: [   ] p: [///]

Acceptance Region
Rejection Region (α)

$p \leq \alpha$
p is inside the Rejection Region

$p > \alpha$
p extends into the Acceptance Region

- Center graph: **if** the hatched area representing p is entirely in the shaded Rejection Region (because $p \leq \alpha$) we **Reject the Null Hypothesis.**
- Right graph: **If** p extends into the white Acceptance Region (because $p > \alpha$), **we Accept (or "Fail to Reject") the Null Hypothesis.**

For example, here is a portion of the output from an analysis which includes an F-test. $\alpha = 0.05$.

Factors	F	Effect Size	p
A: Detergent	729	6.75	0.02
B: Water Temp.	225	3.74	0.04
C: Washing Machine	49	1.75	0.09

- We see that $p < \alpha$ for both Factor A and Factor B. So, we can say that A and B <u>do</u> have a Statistically Significant effect. (We Reject the Null Hypothesis.)
- The p-value for A is considerably smaller than that for B, so the evidence is stronger that A has an effect.
- $p > \alpha$ for Factor C, so we conclude that C does <u>not</u> have a Statistically Significant effect. (We Accept/Fail to Reject the Null Hypothesis.)

 5. Alpha defines the Critical Value(s) of Test Statistics, such as z, t, F, or Chi-Square. The Critical Value or Values, in turn, define the Confidence Interval.

We explained how Alpha plays a key role in the Hypothesis Testing method of Inferential Statistics. It is also an integral part of the other main method – Confidence Intervals. This is explained in detail in the article, *Confidence Intervals – Part 1*. It is also illustrated in the following concept flow diagram (follow the arrows):

Here's how it works. Let's say we want a Confidence Interval around the Mean height of males.

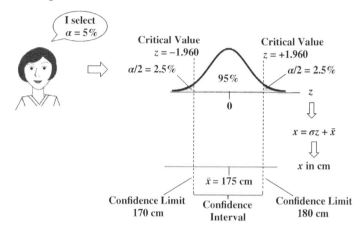

Top part of the diagram:

- The person performing the analysis selects a value for Alpha.
- Alpha – split into two halves – is shown as the shaded areas under the two tails of the curve of a Test Statistic, like z.
- Tables or calculations provide the values of the Test Statistic which form the boundaries of these shaded $\alpha/2$ areas. In this example, $z = -1.960$ and $z = +1.960$.
- These values are the Critical Values of the Test Statistic for $\alpha = 5\%$. They are in the units of the Test Statistic (z is in units of Standard Deviations).

Bottom part of the diagram:

- A Sample of data is collected and a Statistic (e.g., the Sample Mean, $\bar{x}$) is calculated (175 cm in this example).
- To make use of the Critical Values in the real world, we need to convert the Test Statistic Values into real-world values – like centimeters in the example above.

There are different conversion formulas for different Test Statistics and different tests. In this illustration, z is the Test Statistic and it is defined as $z = (x - \bar{x})/\sigma$. So $x = \sigma z + \bar{x}$. We multiply σ (the Population Standard

Deviation), by each critical value of z (-1.960 and $+1.960$), and we add those to the Sample Mean (175 cm).

- That converts the Critical Values -1.960 and $+1.960$ into the Confidence Limits of 170 and 180 cm.
- These Confidence Limits define the lower and upper boundaries of the Confidence Interval.

To further your understanding of how Alpha is used, it would be a good idea to next read the article *Alpha, p, Critical Value, and Test Statistic – How they Work Together.*

Related Articles in This Book: *Alpha and Beta Errors*; *p, p-Value*; *Statistically Significant*; *Alpha, p, Critical Value, and Test Statistic – How They Work Together*; *Test Statistic*; *p, t, and F: ">" or "<"?*; *Hypothesis Testing – Part 1: Overview*; *Critical Value*; *Confidence Intervals – Parts 1 and 2*; *z*

ALPHA AND BETA ERRORS

Summary of Keys to Understanding

 1. There is a risk of an Alpha (aka Type I) **Error or a Beta** (aka Type II) **Error in any Inferential Statistical analysis.**

2.	Alpha Error, "False Positive"	Beta Error, "False Negative"
	I saw a unicorn.	Smoking doesn't cause cancer.
What it is	The error of <u>concluding that there is something</u> – a difference, or a change, or an effect – **when, in reality, there is not.**	The error of <u>concluding that there is nothing</u> – no difference, or no change, or no effect, **when, in reality, there is.**
In Hypothesis Testing	The error of Rejecting the Null Hypothesis when it is true.	The error of Failing to Reject the Null Hypothesis when it is false.
Found in:	Hypothesis Testing and Confidence Levels, *t*-tests, ANOVA, ANOM, etc.	

 3. There is a tradeoff between Alpha and Beta Errors.

The subject being analyzed determines which type is more troublesome.

 4. To <u>reduce</u> both Alpha and Beta Errors, <u>increase</u> the Sample Size.

Explanation

| 1. There is a risk of an **Alpha** (aka Type I) **Error or a Beta** (aka Type II) **Error in any Inferential Statistical analysis.** |

2.	**Alpha Error (False Positive)**	**Beta Error (False Negative)**
	I saw a unicorn.	Smoking doesn't cause cancer.
What it is	The error of <u>concluding that there is something</u> – a difference, or a change, or an effect – **when, in reality, there is not.**	The error of <u>concluding that there is nothing</u> – no difference, or no change, or no effect – **when, in reality, there is.**
In Hypothesis Testing	The error of Rejecting the Null Hypothesis when it is true.	The error of Failing to Reject the Null Hypothesis when it is false.
Also known as	Type I Error, Error of the First Kind Colloquially: **False Positive**, False Alarm, Crying Wolf	Type II Error, Error of the Second Kind, **False Negative**
Found in:	Hypothesis Testing and Confidence Levels, *t*-tests, ANOVA, ANOM, etc.	
Example: in blood tests	Indicate a disease in a healthy person.	Fail to find a disease that exists.
Probability of the error	p	β (Beta)

In Descriptive Statistics, we have complete data on the entire universe we wish to observe. So we can just directly calculate various properties like the Mean or Standard Deviation.

On the other hand, in **Inferential Statistics** methods like Hypothesis Testing and Confidence Intervals, **we don't have the complete data**. The Population or Process is too big or it is always changing, so we can never be 100% sure about it. **We can collect a Sample of data and make an**

estimate from that. As a result, **there will always be a chance for error**. There are two types of this kind of Sampling Error; they are like mirror images of each other.

It may be easiest to think in terms of "False Positive" and "False Negative."

False Positive (Alpha Error) – **is the error of concluding that there is a difference, change, or effect, when, in reality there is no difference, change, or effect.**

"False Negative" is the opposite – **the error of concluding there is nothing happening, when, in fact, something is.** For example, the statistical analysis of a Process Mean concluded that it has not changed over time, when, in reality the Process Mean has "drifted."

In this context **"positive" does not mean "beneficial," and "negative" does not mean "undesirable."** In fact, for medical diagnostic tests, a "positive" result indicates that a disease was found. And a "negative" result is no disease found.

Alpha, α (*see the article by that name*) **is selected by the tester as the maximum Probability of an Alpha** (aka Type 1 aka False Positive) **Error they will accept** and still be able to call the results "Statistically Significant." That's why Alpha is called the "Significance Level" or "Level of Significance."

Beta, β, is the Probability of a Beta Error. Unlike Alpha, which is selected by us, Beta is calculated by the analysis. $1 - \beta$ is the Probability of there not being a Beta Error. So, if we call Beta the Probability of a False Negative, we might think of $1 - \beta$ as the Probability of a "true negative." $1 - \beta$ is called the "Power" of the test, and it is used in Design of Experiments to determine the required Sample Size.

You may have noticed a lack of symmetry in the terminology. This can be confusing; hopefully the following table will help:

p is the Probability of an Alpha Error	β is the Probability of a Beta Error
α is the maximum acceptable Probability for an Alpha Error	
$1 - \alpha$ is called the Confidence Level	$1 - \beta$ is called the Power of the test

In Hypothesis Testing

Let's say we're testing the effect of a new medicine compared to a placebo. **The Null Hypothesis (H$_0$) says that there is no difference** between the new medicine and the placebo.

- **If the reality is that there is no difference (H_0 is true), and if …**
  - our testing concludes that there is no difference, then there is no error.
  - **our testing concludes that there is a difference, then there is an Alpha Error.**
- **If the reality is that there is a difference (H_0 is false), and if …**
  - **our testing concludes that there is no difference, then there is a Beta Error**
  - our testing concludes that there is a difference, then there is no error.

Conclusion from our testing	Reality: No difference, H_0 is True	Reality: There is a difference, H_0 is False
Accept (Fail to Reject) H_0	No error	Beta Error
Reject H_0	Alpha Error	No error

 | **3. There is a tradeoff between Alpha and Beta Errors.** |

This makes sense. Consider the situation of airport security scanning. We want to detect metal weapons. We don't adjust the scanner to detect only metallic objects which are the size of an average gun or knife or larger. That would **reduce the risk of Alpha Errors** (e.g., identifying coins as possible weapons), but it **would increase the risk of Beta Errors** (not detecting small guns and knives).

This is the reason why we don't select an Alpha (maximum tolerable Probability of an Alpha Error) which is much smaller than the usual 0.05. There is a price to pay for making α extremely small. And the price is making the Probability of a Beta Error larger.

So, we need to select a value for Alpha which balances the need to avoid both types of error. The consensus seems to be that 0.05 is good for most uses.

How to make the tradeoff between Alpha and Beta depends on the situation being analyzed. In some cases, the effect of an Alpha Error is relatively benign and you don't want to risk a False Negative. In other cases, the opposite is true. Some examples:

Situation	Consequence of an Alpha Error (False Positive)	Consequence of a Beta Error (False Negative)	Wise choice for level of risk	
			Alpha Error (risk of False Positive)	Beta Error (risk of False Negative)
Airport Security	Detain an innocent person as a terrorist	Let a terrorist on board	higher	lower
Inspect critical components for jet engine	Reject a good component	Engine failure	higher	lower
Inspect painting on the underside of a wheelbarrow	Cost of a reject	Customer will probably not notice or care	lower	higher

 4. To reduce both Alpha and Beta Errors, increase the Sample Size.

There are other factors involved, but increasing the Sample Size will reduce both Alpha and Beta Errors. If the Sample Size is relatively large, increasing it further will yield diminishing returns in error reduction. (See the articles on Sample Size.)

Related Articles in This Book: *Alpha, α*; *Alpha, p-Value, Critical Value, and Test Statistic – How They Work Together*; *p, p-Value*; *Inferential Statistics*; *Power*; *Sample Size – Parts 1 and 2*

ALPHA, p, CRITICAL VALUE, AND TEST STATISTIC – HOW THEY WORK TOGETHER

Summary of Keys to Understanding

 1. **Alpha and p are Cumulative Probabilities.** They are represented as **areas under the curve** of the Test Statistic Distribution.

 2. **The Critical Value** (e.g., z-critical) **and the value of the Test Statistic** (e.g., z) **are point values** on the horizontal axis of the Test Statistic Distribution. **They mark the inner boundaries of the areas representing Alpha and p, respectively.**

 3. **The person performing the analysis selects the value of Alpha, α.**
    Alpha and the Distribution are then used to calculate the Critical Value of the Test Statistic (e.g., z-critical). It is the value which forms the inner boundary of Alpha.

 4. **Sample data are used to calculate the value of the Test Statistic** (e.g., z).
    The value of the Test Statistic and the Distribution are then used to calculate the value of p. p is the area under the curve outward from this calculated value of the Test Statistic.

Areas under the curve (right tail)

α: ☐ p: ▨

Statistically Significant:
Reject H_0

z-critical z

$z \geq z$-critical, so $p \leq \alpha$

Not Statistically Significant:
Fail to Reject H_0

z z-critical

$z < z$-critical so $p > \alpha$

 5. To determine Statistical Significance, compare p to Alpha, or (equivalently) compare the value of the Test Statistic to its Critical value.

If $p \leq \alpha$ or (same thing) $z \geq z$-critical,

then there is a Statistically Significant difference, change, or effect. Reject the Null Hypothesis, H_0.

Explanation

Much of statistics involves taking a Sample of data and using it to infer something about the Population or Process from which the Sample was collected. This is called Inferential Statistics.

There are 4 key concepts at the heart of Inferential Statistics:

- Alpha, the Level of Significance
- *p*, the Probability of an Alpha (False Positive) Error
- a Test Statistic, such as z, t, F, or χ^2 (and its associated Distribution)
- Critical Value, the value of the Test Statistic corresponding to Alpha

This article describes how these 4 concepts work together in Inferential Statistics. It assumes you are familiar with the individual concepts. If you are not, it's easy enough to get familiar with them by reading the individual articles for each of them.

	Alpha, α	**p**	**Critical Value of Test Statistic**	**Test Statistic Value**
What is it? How is it pictured?	a Cumulative Probability		a value of the Test Statistic	
	an area under the curve of the Distribution of the Test Statistic		a point on the horizontal axis of the Distribution of the Test Statistic	
Boundary	Critical Value marks its boundary	Test Statistic Value marks its boundary	Forms the boundary for Alpha	Forms the boundary for *p*
How is its value determined?	Selected by the tester	area bounded by the Test Statistic value	boundary of the Alpha area	calculated from Sample Data
Compared with	*p*	α	Test Statistic Value	Critical Value of Test Statistic
Statistically Significant/ Reject the Null Hypothesis if	$p \leq \alpha$		Test Statistic $\geq$ Critical Value e.g., $z \geq z$-critical	

The preceding compare-and-contrast table is a visual summary of the 5 Keys to Understanding from the previous page and the interrelationships among the 4 concepts. This article will cover its content in detail. At the end of the article is a concept flow visual which explains the same things as this table, but using a different format. Use whichever one works better for you.

> **1. Alpha and *p* are Cumulative Probabilities.** They are represented as **areas under the curve** of the Test Statistic Distribution.

A Test Statistic is calculated using Sample data. But, unlike other Statistics (e.g., the Mean or Standard Deviation), Test Statistics have an associated Probability Distribution (or family of such Distributions). Common Test Statistics are z, t, F, and χ^2 (Chi-Square).

The Distribution is plotted as a curve over a horizontal axis. The Test Statistic values are along the horizontal axis. The Point Probability of any value of a Test Statistic is the height of the curve above that Test Statistic value. But, we're really interested in Cumulative Probabilities.

A Cumulative Probability is the total Probability of all values in a range. Pictorially, **it is shown as the area under the part of curve of the Distribution which is above the range.**

In the diagram below, the curve of the Probability Distribution is divided by x into two ranges: negative infinity to x and x to infinity. Above these two ranges are two areas (unshaded and shaded) representing two Cumulative Probabilities. The total area of the two is 100%.

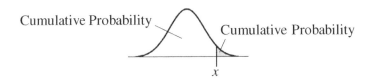

In calculus-speak, the area under a curve is calculated as the integral of the curve over the range. Fortunately, when we use Test Statistics, we don't have to worry about calculus and integrals. The areas for specific values of the Test Statistic are shown in tables in books and websites, or they can be calculated with software, spreadsheets, or calculators on websites.

For example, if we select Alpha to be 5% (0.05), and we are using the Test Statistic z, then the value of z which corresponds to that value of

Alpha is $z = 1.645$

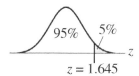

95%   5%

$z = 1.645$

 2. The Critical Value (e.g., *z*-critical) **and the value of the Test Statistic** (e.g., *z*) **are point values** on the horizontal axis of the Test Statistic Distribution. **They mark the inner boundaries of the areas representing Alpha and *p*, respectively.**

- **The Critical Value is determined from the Distribution of the Test Statistic and the selected value of Alpha.** For example, as we showed earlier, if we select $\alpha = 5\%$ and we use *z* as our Test Statistic, then *z*-critical $= 1.645$.

- **The Sample data are used to calculate a value of the test Statistic.** For example, the following formula is used to calculate the value of *z* from Sample data:

$$z = (\mu - \bar{x})/s$$

where $\bar{x}$ is the Sample Mean, *s* is the Sample Standard Variation, and μ is a specified value, for example, a target or historical value for the Mean.

The following tables illustrate some values for a 1-tailed/right-tailed situation (only shading under the right tail. See the article "*Alpha, α*" for more on 1-tailed and 2-tailed analyses.) Notice that **the larger the value of the boundary**, the farther out it is in the direction of the tail, and so **the smaller the area under the curve.**

As the boundary point value grows larger ——————————>

Boundary: *z* or *z*-critical	0	0.675	1.282	1.645	2.327
Area: *p* or *α*	0.50	0.25	0.1	0.05	0.01

<————————— **the Cumulative Probability area grows smaller**

The graphs below are close-ups of the right tail of the *z* Distribution. The shaded area represents the Cumulative Probability, Alpha. The hatched area represents the Cumulative Probability, *p*. As explained in the tables above, **the larger the point value (*z* or *z*-critical), the smaller the value for its corresponding Cumulative Probability (*p* or *α*, respectively).**

Areas under the curve (right tail)

α: ☐ p: ▨

z-critical z z z-critical

$z \geq z$-critical, so $p \leq \alpha$ $z < z$-critical so $p > \alpha$

The left diagram above shows a value of z which is greater (farther from the Mean) than the Critical Value. So, p, the area under the curve bounded by z, is smaller than the area for Alpha, which is bounded by the Critical Value. The right diagram shows the opposite.

> **3. The person performing the analysis selects the value of Alpha.**
>
>   **Alpha and the Distribution are then used to calculate the Critical Value of the Test Statistic** (e.g., z-critical). It is the value which forms the inner boundary of Alpha.

Alpha is called the Level of Significance. **Alpha is the upper limit for the Probability of an Alpha/"False-Positive" Error below which any observed difference, change, or effect is deemed Statistically Significant.** This is the only one of the four concepts featured in this article which is not calculated. It is selected by the person doing the analysis. Most commonly, $\alpha = 5\%$ (0.05) is selected. This gives a Level of Confidence of $1 - \alpha = 95\%$.

If we then plot this as a shaded area under the curve, the boundary can be calculated from it.

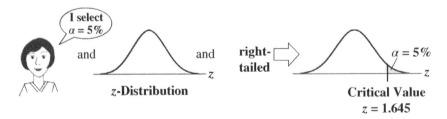

Note: for a 2-tailed analysis, half of Alpha (2.5%) would be placed under each tail. (left and right. The Critical Value would be 1.96 on the right and – 1.96 on the left).

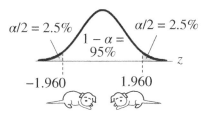

> **4. Sample data are used to calculate the value of the Test Statistic** (e.g., z).
>
> **The value of the Test Statistic and the Distribution are then used to calculate the value of *p*.** p is the area under the curve outward from this calculated value of the Test Statistic.

We saw how we use a Cumulative Probability (α) to get a point value (the Critical Value). We'll now go in the opposite direction. We use a point value for the Test Statistic, z, to get a Cumulative Probability (p).

p **is the <u>actual</u> Probability of an Alpha Error** for a particular Sample of data.

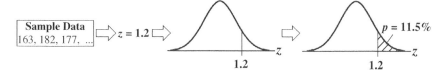

Let's say that z is calculated from the Sample data to be 1.2. This gives us a value of $p = 0.115$ (11.5%).

> **5. To determine Statistical Significance, compare *p* to Alpha,** or (equivalently) **compare the value of the Test Statistic to its Critical value.**
>
> **If $p \leq \alpha$** or (same thing) $z \geq z$**-critical,**
>
> **then there <u>is</u> a Statistically Significant difference, change, or effect. Reject the Null Hypothesis (H_0).**

- We selected Alpha as the Level of Significance – the maximum Probability of an Alpha/"False-Positive" Error) which we are willing to tolerate.
- We calculated p as the actual Probability of an Alpha Error for our Sample.

- **So if $p \leq \alpha$, then any difference, change, or effect observed in the Sample data is Statistically Significant.**

Note that:

- **The Critical Value is determined from Alpha. The two contain the same information**. Given a value for one, we could determine the other from the Distribution.
- Similarly, **the value of the Test Statistic (z in our example) contains the same information as p.**
- **So, comparing the Test Statistic to the Critical Value of the Test Statistic is statistically identical to comparing p to Alpha.**

Therefore:

If $p \leq \alpha$ or $z \geq z$-critical,

then there is a Statistically Significant difference, change, or effect.

(Reject the Null Hypothesis).

Depicted graphically:

<u>Areas under the curve (right tail)</u>

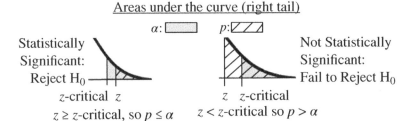

The table at the beginning of this article summarized the whole article in one visual. On the next page is the same information presented in another way. Use whichever one works best for you.

t is the Test Statistic in this illustration.

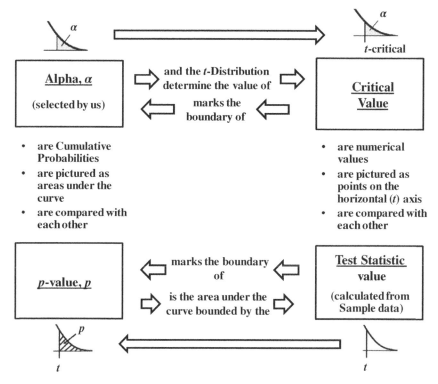

Related Articles in This Book: *Alpha, α; p-Value, p; Critical Value; Test Statistic; Distributions – Part 1: What They Are; Inferential Statistics; Hypothesis Testing – Parts 1–3; Confidence Intervals – Parts 1 and 2; p, t, and F: "<" or ">"?*

ALTERNATIVE HYPOTHESIS

Recommendation: read the article "Null Hypothesis" before reading this article.

Symbols for the Alternative Hypothesis: H_A, H_1, or H_a

Summary of Keys to Understanding

 1. Stating a **Null Hypothesis** (H_0) **and an Alternative Hypothesis** (H_A) **is the first step in our 5-step method for Hypothesis Testing.**

 2. **The Alternative Hypothesis is the opposite of the Null Hypothesis – and vice versa.**

 3. **Stating the Alternative Hypothesis as a comparison formula,** rather than in words, **can make things easier to understand. The formula must include an inequivalence in the comparison operator,** using one of these: "$\neq$", "$>$", or "$<$".

Comparison Operator		Tails of the Test	
H_A	H_0		
$\neq$	$=$	2-tailed	$\alpha/2$ $\alpha/2$
$>$	$\leq$	Right-tailed	$\alpha = 5\%$
$<$	$\geq$	Left-tailed	$\alpha = 5\%$

 4. **In a 1-tailed test, the Alternative Hypothesis** (aka the "Research Hypothesis" or Maintained Hypothesis") **tells you in which direction** (right or left) **the tail points.**

Explanation

> **1.** Stating a **Null Hypothesis** (H_0) **and an Alternative Hypothesis** (H_A) **is the first step in our 5-step method for Hypothesis Testing.**

Hypothesis Testing is one of two common methods for Inferential Statistics. Confidence Intervals is the other. In Inferential Statistics, we estimate a statistical property (e.g., the Mean or Standard Deviation) of a Population or Process by taking a Sample of data and calculating the property in the Sample.

In the article, *"Hypothesis Testing – Part 2: How To"* we describe a 5-step method of Hypothesis Testing:

1. State the problem or question in the form of a Null Hypothesis (H_0) and Alternative Hypothesis (H_A).
2. Select a Level of Significance (α).
3. Collect a Sample of data for analysis.
4. Perform a statistical analysis on the Sample data.
5. Come to a conclusion about the Null Hypothesis (Reject or Fail to Reject).

Hypothesis Testing can be very confusing, mainly because the language in steps 1 and 5 can be confusing. This article and the *Null Hypothesis* article are written to clear up the confusion in step 1.

Experts disagree on whether an Alternative Hypothesis should be used. It is included here, because, as we'll explain later, it is useful in 1-tailed tests.

> **2. The Alternative Hypothesis is the opposite of the Null Hypothesis – and vice versa.**

What exactly does that mean? It means that:

- **If the Null Hypothesis is true, then the Alternative Hypothesis is false.**
- **If the Null Hypothesis is false, then the Alternative Hypothesis is true.**

These two statements imply that:

H_0 **and** H_A **are**

- **mutually exclusive and**
- **collectively exhaustive.**

This means that either H_0 or H_A must be true; you can't have neither being true. And you can't have both being true.

Here are some examples:

Example 1

H_0: There is no difference between the Standard Deviations of Population A and Population B.

H_A: There is a difference between the Standard Deviations of Population A and Population B.

Example 2

H_A: Our school's average test scores are better than the national average.

H_0: Our school's average test scores are less than or equal to the national average.

Example 3

H_A: Our orders ship in less than 4 days.

H_0: Our orders ship in 4 days or more

In addition to being mutually exclusive and collectively exhaustive, these three examples include a couple of other concepts:

- **Statistically Significant:** In Example 1, our two Samples of data will no doubt show some difference in the two Standard Deviations. The Inferential Statistical test will determine whether that difference is Statistically Significant. Likewise, the "better than" and "less than" in Examples 2 and 3 are implicitly modified by "to a Statistically Significant extent."

- **2-tailed or 1-tailed:** As we'll explain later, Example 1 will use a 2-tailed analysis. Example 2 (right-tailed) and Example 3 (left-tailed) are 1-tailed.

Note also that for Examples 2 and 3, we list H_A first and H_0 second. The reason for this is explained below, under Keys to Understanding #4.

 | **3. Stating the Alternative Hypothesis as a comparison formula**, rather than in words, **can make things easier to understand. The formula must include an inequivalence in the comparison operator**, using one of these: "$\neq$", "$>$", or "$<$".

Null and Alternative Hypotheses involve comparisons (equations or inequalities) between values of Parameters (properties) of Populations or

Processes. A Parameter could be a Mean (μ), a Standard Deviation (σ), or other descriptive statistical property.

In a Hypothesis, a Parameter from one Population or Process could be compared with that of another, for example,

$$\sigma_A = \sigma_B$$

Or it could be compared with a numerical value, like a target or historical value:

$$\sigma < 1.5$$

There are 3 basic comparison symbols: equal "=", greater than ">", and less than "<".

There are also compound symbols: not equal $\neq$, greater than or equal to "$\geq$", and less than or equal to "$\leq$".

Comparison Operator		Tails of the Test	
H_A	H_0		
$\neq$	$=$	2-tailed	$\alpha/2$ $\alpha/2$
$>$	$\leq$	Right-tailed	$\alpha = 5\%$
$<$	$\geq$	Left-tailed	$\alpha = 5\%$

 4. In a 1-tailed test, the Alternative Hypothesis (aka the "Research Hypothesis" or Maintained Hypothesis") **tells you in which direction** (right or left) **the tail points.**

If H_0 can be stated with an equal sign, "=", the situation is relatively straightforward. We are only interested in whether there is a Statistically Significant difference, change, or effect. There is no direction involved. When we tell our statistical tool what type of test it is, we say "2-tailed." The common wisdom is to state a Null Hypothesis, and then the Alternative Hypothesis is the opposite.

But, for 1-tailed tests, when "<" or ">" is involved, it gets more complicated. Once we determine which is the Null and which is the Alternative Hypothesis, it's easy to assign a comparison operator to each comparison formula. **But how do we decide which is which?**

It may help to know that **the Alternative Hypothesis is also known as the Research Hypothesis or the Maintained Hypothesis.** And that **the Alternative Hypothesis is the one that the researcher maintains and aims to prove.**

In Example 2 above, our school's average test scores are somewhat better than the national average, and we would like to prove that this is a Statistically Significant difference. So we select as our Alternative (Maintained) Hypothesis:

$$H_A: \mu_{school} > \mu_{national}$$

The Null Hypothesis then becomes:

$$H_0: \mu_{school} \leq \mu_{national}$$

Note that, **for 1-tailed tests, it is better to start with a statement of the Alternative Hypothesis and then derive the Null Hypothesis as the opposite.** This is because we know what we maintain and would like to prove.

Furthermore, **the "<" or ">" in the Alternative Hypothesis points in the direction of the tail.** "<" in the Alternative Hypothesis means that the test is Left-Tailed. ">" tells us that it is Right-Tailed.

Related Articles in This Book: *Null Hypothesis*; *Hypothesis Testing – Part 1: Overview*; *Hypothesis Testing – Part 2: How To*; *Reject the Null Hypothesis*; *Fail to Reject the Null Hypothesis*

ANALYSIS OF MEANS (ANOM)

Summary of Keys to Understanding

1. Analysis of Means (ANOM) tells us whether the Means from several Samples are statistically the same as the Overall Mean.

2. ANOM has some similarities to, and some differences from, ANOVA

	ANOM	ANOVA
Assumptions	Approximately Normal data	
Analyzes Variation of several Means	Yes	
1-Way or 2-Way	Yes	
Variation	around the overall Mean	among each other
Identifies which Means are not statistically the same	Yes	No
Output	Graphical	Statistical: ANOVA Table

3. The graphical ANOM output is similar to a Control Chart.

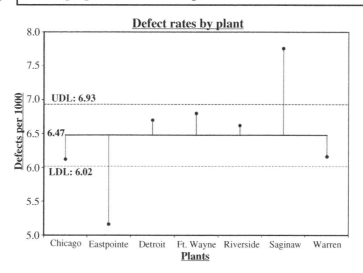

Explanation

 1. **Analysis of Means (ANOM) tells us whether the Means from Samples from several different Populations or Processes are statistically the same as the Overall Mean.**

The different Populations or Processes are represented by different values of a Categorical/Nominal Variable. As such, they are names, for example,

Call center reps: John, Jane, Robert, Melissa, Judith, Mike

Vendors: Company A, Company B, Company C, Company D

Plants: Chicago, Eastpointe, Detroit, Fort Wayne, Riverside, Toledo, Warren

The Means here are the Means of an Independent Variable, y. y is numerical, such as the number of calls successfully handled, delivery times, and defect rates.

For each name, there will be a Sample of data – for example, for each call center rep, the number of calls handled each day for a number of days.

The Overall Mean, sometimes called the Grand Mean, is the average of all the y-Variable values from all the Samples.

ANOM has been most frequently used in industrial and process-improvement analyses, but it is applicable generally.

The underlying calculations for ANOM are more complicated than those for ANOVA, and explaining them is beyond the scope of this book.

 2. **ANOM has some similarities to, and some differences from, ANOVA**

	ANOM	ANOVA
Assumptions	Approximately Normal data	
Analyzes Variation of several Means	Yes	
1-Way or 2-Way	Yes	

First, the similarities: In order to produce valid results, **both ANOM and ANOVA require that the data be approximately Normal.** "Approximately" Normal is not strictly defined, but the data should not be obviously non-Normal. That is, it should have one discernable peak and not be strongly skewed.

Second, they both analyze Variation in Means. ANOVA is "Analysis of Variation," but it analyzes Variation among Means. Both are usually used with 3 or more Means. For 2 Means, there is the 2-Sample t-test.

And both can perform 1-Way (aka Single Factor, i.e., one *x* Variable) or 2-Way (Two Factor, two *x* Variables) analyses.

	ANOM	ANOVA
Variation	around the overall Mean	among each other
Identifies which Means are not statistically the same	Yes	No

ANOM calculates the Overall Mean, and then it measures the Variation of each Mean from that. In the conceptual diagram below, each Sample is depicted by a Normal curve. The distance between each Sample Mean and the Overall Mean is identified as a "Variation."

ANOM retains the identity of the source of each of these Variations (#1, #2, and #3), and it displays this graphically in the ANOM chart (shown later in this article).

ANOM

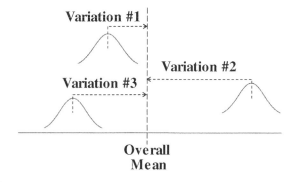

ANOVA

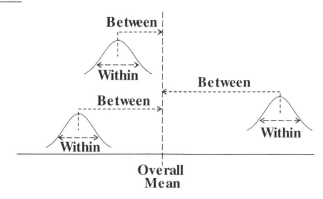

ANOVA takes a more holistic approach, in which the identity of the individual Sample Variations is lost. This is explained in detail in the articles, *ANOVA, Parts 1, 2, and 3.* But briefly, …

ANOVA starts out like ANOM, calculating the Variation between each Sample Mean and the Overall Mean. But it then consolidates this information into one Statistic for all the Samples, the Mean Sum of Squares Between, MSB.

Next it calculates Variation within each Sample and then consolidates that into one Statistic, the Mean Sum of Squares Within, MSW*. So **any information about individual Sample Means and Variances is lost.** That is why **ANOVA can only tell us if there is a Statistically Significant difference somewhere among the Means, not which one(s) are Significantly different. However, ANOM can.**

*(ANOVA goes on to divide MSB by MSW, yielding the F-statistic, which is then compared to F-critical to determine Statistical Significance.)

 3. The graphical ANOM output is similar to a Control Chart.

The output from ANOVA is a table of Statistics. **The output from ANOM is graphical.**

Example: Let's say we have 7 plants mass-producing the same product, and we want to determine whether any have a defect rate per thousand which is (Statistically) Significantly better or worse than the others. We collect data for 5 days.

	Chicago	Eastpointe	Detroit	Ft. Wayne	Riverside	Saginaw	Warren
	6.0	5.2	6.8	7.1	6.8	7.4	6.2
	6.5	4.3	7.0	6.7	6.0	7.9	6.9
	6.1	5.1	6.7	6.5	6.4	8.2	5.9
	6.2	5.3	6.4	6.9	7.3	7.7	5.7
	5.8	5.9	6.6	6.8	6.6	7.6	6.1
Means:	6.1	5.2	6.7	6.8	6.6	7.8	6.2

In the ANOM chart below, the dotted horizontal lines, the **Upper Decision Line (UDL) and Lower Decision Line (LDL) define a Confidence Interval**, in this case, for $\alpha = 0.05$. Our conclusion is that only Eastpointe (on the low side) and Saginaw (on the high side) exhibit a Statistically Significant difference in their Mean defect rates. So **ANOM tells us not only whether any plants are Significantly different, but also which ones are.**

ANOM Output The dots show the Means of the 5 days of data for each plant.

The Overall Mean for all plants is 6.47.

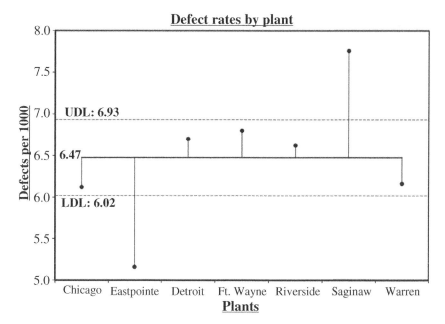

Related Articles in This Book: *ANOVA, Parts 1–4*; *Variation/Variability/Dispersion/Spread*; *Confidence Intervals – Parts 1 and 2*; *Alpha, a*; *Control Charts – Part 1: General Concepts and Principles*

ANOVA – PART 1 (OF 4): WHAT IT DOES

Summary of Keys to Understanding

1. **"ANOVA" is an acronym for ANalysis Of VAriance. However, its objective is to determine if one or more of the Means of several Groups is different from the others.**

2. **Assumptions (test requirements) are**
   - The groups being compared have a roughly Normal Distribution
   - The groups have similar Variances

3. **There are 3 types of ANOVA**
   - 1-Way aka Single Factor
   - 2-Way without Replication
   - 2-Way with Replication

4. **ANOVA is often used in Designed Experiments. An ANOVA Table is often an output in Multiple Linear Regression analysis.**

5. ANOVA <u>Does</u>	ANOVA Does <u>Not</u>		Do this instead
compare several Means with <u>each other</u>	compare several Means with <u>the overall Mean</u>		ANOM
say <u>whether or not</u> there is a difference among Means	say <u>which</u> Means differ		ANOM or Confidence Intervals
require Continuous data	handle Discrete data		Chi-square Test of Variance
require roughly Normal Distributions	handle very Non-Normal Distributions		Kruskal–Wallis
require somewhat equal Sample Variances	handle very unequal Sample Sizes and Variances		Ensure equal Sample Sizes when Sample Variances are unequal

Explanation

There are 4 articles in this series about ANOVA

Part 1: What it Does

Part 2: How it Does It

The underlying 7-step method which involves Sums of Squares and an *F*-test. Students may need to understand this for their exams. But if you just want the answer to the ANOVA analysis, spreadsheets or software can give it to you if you just provide the data.

Part 3: 1-Way

The method used when there is a single Factor affecting the outcome we are measuring. For example, the single Factor would be the drug used in a test. ANOVA would be used to determine whether any stood out from the rest.

Part 4: 2-Way

Used when 2 Factors affect the outcome. For example, in a laundry process, measuring the effect on cleanliness of the Factors, water temperature, and detergent type. Interactions between Factors are an important component of 2-Way ANOVA.

 1. "ANOVA" is an acronym for <u>A</u>Nalysis <u>Of</u> <u>VA</u>riance. However, its objective is to determine if one or more of the Means of several Groups are different from the others.

ANOVA is an acronym for "Analysis of Variance." But **analyzing Variances is not its objective**. **Its objective is to determine whether one or more of several Means are different** from the others by a Statistically Significant amount. **It does this by analyzing Variances.**

"Group" here is a generic term which can refer to:

– a <u>Population</u> or <u>Process</u> for which we have complete data.

– a <u>Sample</u> taken from a Population or Process, for example, the annual incomes of 30 people who live in particular neighborhood. In the case of a Sample, the Sum of the Squares of the Sample is an estimate of the Sum of the Squares for the <u>Population</u>.

 2. Assumptions (test requirements) are
- Groups being compared have a roughly Normal Distribution
- Groups have similar Variances

As we will see in the Part 2 article, the Variances which are analyzed are not the Variances of the individual groups whose Means we are comparing. The Variances are the Mean Sum of Squares Between groups (MSB) and Mean Sum of Squares Within groups (MSW). These are two numbers, each of which summarizes different information about all the groups.

For ANOVA, the groups should be roughly Normal in their Distributions and their Variances should be roughly similar. ANOVA is fairly tolerant in terms of what is considered Normal enough or having similar enough Variances. If these assumptions are not roughly met, then the Kruskal–Wallis test can be used instead.

> ### 3. There are 3 types of ANOVA
> - 1-Way, aka Single Factor
> - 2-Way without Replication
> - 2-Way with Replication

1-Way, also known as Single Factor, is covered in the Part 3 article. There is one Factor – the x Variable – which affects the outcome, or y Variable. For example, the single Factor could be blood pressure drug. There could be several different drugs being compared. The y Variable would be a measure of reduction in blood pressure.

The 2-Way types of ANOVA are covered in the Part 4 article. In both cases, there are two Factors, or x Variables. For example, water temperature and detergent type would be the two Factors, and a cleanliness measure would be the outcome or y Variables.

If the data show that the two Factors interact, then the 2-Way with Replication (repeated measurements) must be used.

> ### 4. ANOVA is often used in Designed Experiments. An ANOVA Table is often an output in a Multiple Linear Regression analysis.

ANOVA Table					
	df	SS	MS	F	p-value
Regression	−4.000	48,877.931	−12,219.483	32.727	0.009
Residual	3.000	1493.498	497.833		
Total	−1.000	50,371.429			

5. ANOVA Does	ANOVA Does Not		Do this instead
compare several Means with each other	compare several Means with the overall Mean		ANOM
say whether or not there is a difference among Means	say which Means differ		ANOM or Confidence Intervals
require Continuous data	handle Discrete data		Chi-square Test of Variance
require roughly Normal Distributions	handle very Non-Normal Distributions		Kruskal–Wallis
require somewhat equal Sample Variances	handle very unequal Sample Sizes and Variances		Ensure equal Sample Sizes when Sample Variances are unequal

Related Articles in This Book: *Part 2: How It Does It*; *Part 3: 1-Way*; *Part 4: 2-Way*; *Sums of Squares*; *ANOVA vs. Regression*; *Design of Experiments (DOE) – Part 3*; *Regression – Part 4: Multiple Linear*

ANOVA – PART 2 (OF 4): HOW IT DOES IT

Summary of Keys to Understanding

1. **Sum of Squares Within (SSW)** is **the sum of the Variations** (as expressed by the Sums of Squares, SS's) **within each of several Groups.**

$$\text{SSW} = \text{SS}_1 + \text{SS}_2 + \cdots + \text{SS}_n$$

2. **Sum of Squares Between (SSB) measures Variation between** (among) **Groups,**

$$\text{SSB} = \sum n(\overline{X} - \overline{\overline{X}})^2$$

and **Sums of Squares Total (SST) is the Total of both types of Variation.**

$$\text{SST} = \text{SSW} + \text{SSB}$$

3. The Mean Sums (of Squares), MSB and MSW, are averages of SSB and SSW, respectively. **With MSB and MSW, we have only 2 Statistics which summarize the Variation in 3 or more groups.**

4. **Mean Sums of Squares are similar to the Variance. As such, they can be used to calculate the Test Statistic, F,** which is the ratio of two Variances.

5. **Mean Sums of Squares are used in the F-tests in ANOVA. A large value of MSB,** compared with MSW, **indicates that the Sample Means are not close to each other.** This makes for a large value for F, **which makes it more likely that $F \geq F$-critical.**

SSB $\Rightarrow$ MSB

SSW $\Rightarrow$ MSW

$\Rightarrow \dfrac{\text{MSB}}{\text{MSW}} = F \Rightarrow$

If $F \geq F$-critical, $\Rightarrow$ there is a difference.

If $F < F$-critical, $\Rightarrow$ there is no difference.

Explanation

This article is about what goes on behind the scenes in an ANOVA. Spreadsheets or software will do all the calculations for you

The generic **Sum of Squares (SS)** is the **sum of the Squared Deviations** of all the data values in a single Group (e.g., a Sample). **SS is one measure of Variation** (it also happens to be the numerator in the formula for Variance).

$$SS = \sum(x - \bar{x})^2$$

MSB and MSW are special types of Sums of Squares. In this article, we will show how MSB and MSW are derived from the data, starting with the most basic kind of Sum of Squares.

The **Deviation** (of a single data value, *x*) is $x - \bar{x}$.

"Deviation" here means **distance from the Mean**: $x - \bar{x}$, where *x* is an individual data value in a Group, and $\bar{x}$ is the Mean of the Group. It could just as easily be $\bar{x} - x$ as $x - \bar{x}$. For our purposes, we don't care whether a value is less than or greater than the Mean. We just want to know by how much it deviates from the Mean. **So we square it, to ensure we always get a positive number.** (Another article in this book, *Variance,* explains why we don't just use the absolute value instead of squaring).

A **Squared Deviation is just the square of a Deviation.**

If we want to find a measure of Variation for the Group we can total up all the Squared Deviations of all data values in the Sample. That gives us the Sum of the Squared Deviations, aka the **Sum of Squares.**

$$SS = \sum(x - \bar{x})^2$$

So, it is easy to see that – like Variance and Standard Deviation – **Sum of Squares (SS) is a measure of Variation.** In fact, the **Sum of Squares is the numerator in the formula for Variance (s^2).**

$$s^2 = \frac{\sum(x - \bar{x})^2}{n - 1} = \frac{SS}{n - 1}$$

Variance is, for most purposes, a better measure of Variation than the generic SS, because it takes into account the Sample Size, and it approximates the square of the average Deviation. But there is more to the SS story. **ANOVA uses 3 particular types of Sums of Squares: Within, Between, and Total (SSW, SSB, and SST).** Whereas the generic SS is only about a single Group, **these three each measure different kinds of Variation involving multiple Groups.**

> 1. **Sum of Squares Within (SSW)** is **the sum of the Variations** (as expressed by the Sums of Squares, SS's) **within each** of **several Groups.**
>
> $$SSW = SS_1 + SS_2 + \cdots + SS_n$$

Sums of Squares Within (SSW) summarizes how much Variation there is within each of several Groups (usually Samples) – by giving the sum of all such **Variations**.

This is not numerically precise, but conceptually, one might picture SS as the width of the "meaty" part of a Distribution curve – the part without the skinny tails on either side.

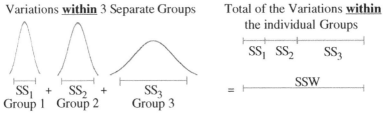

Variations **within** 3 Separate Groups

SS_1 + SS_2 + SS_3
Group 1 Group 2 Group 3

Total of the Variations **within** the individual Groups

SS_1 SS_2 SS_3

= SSW

A comparatively small SSW indicates that the data within the individual Groups are tightly clustered about their respective Means. If the data in each Group represent the effects of a particular treatment, for example, this **is indicative of consistent results** (good or bad) within each individual treatment.

"Small" is a relative term, so the word **"comparatively" is key here.** We'll need to compare SSW with another type of Sum of Squares (SSB) before being able to make a final determination.

A comparatively large SSW shows that the **data** within the individual Groups **are widely dispersed. This would indicate inconsistent results** within each individual treatment.

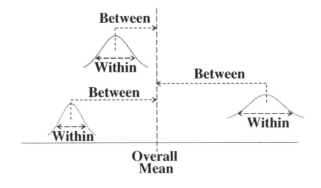

Between
Within
Between
Between
Within
Within

Overall
Mean

> **2. Sum of Squares Between (SSB) measures Variation between** (among) **Groups,**
>
> $$SSB = \sum n(\overline{X} - \overline{\overline{X}})^2$$
>
> and **Sums of Squares Total (SST) is the Total of both types of Variation.**
>
> $$SST = SSW + SSB$$

(According to the dictionary, "between" is about 2 things, so the word is used ungrammatically here; it should be "among." However, "between" is much more commonly used in this context, so we'll go with that in this book.)

To calculate Sum of Squares between, SSB:

$$SSB = \sum n(\overline{X} - \overline{\overline{X}})^2$$

where $\overline{X}$ is a Group Mean and $\overline{\overline{X}}$ is the Overall Mean and n is the number of values in that Group. The Overall Mean (also called the Grand Mean) is the Mean of all the data values from all Groups.

– First, calculate the Overall Mean, (symbols $\overline{\overline{X}}$). You can forget the individual groupings, just add up the data values from all Groups and divide by the total number of values.

  In the form of a formula – Overall Mean: $\overline{\overline{X}} = \sum x_{ij}/N$

  where i represents the individual values in one Sample

  and j represents the individual Samples

  and N is the total of Sample Sizes of all Samples

  For example, Sample #1 has values 40, 45, 45, 50, and a Mean of 45; Sample #2 has values 25, 35, 35, 45, and a Mean of 35; Sample #3 has values 40, 55, 55, 70, and a Mean of 55.

$$\overline{\overline{X}} = \frac{40 + 45 + 45 + 50 + 25 + 35 + 35 + 45 + 40 + 55 + 55 + 70}{12}$$

$$= \frac{540}{12} = 45$$

– Next, subtract the Overall Mean, $\overline{\overline{X}}$, from each Group Mean $\overline{X}_j$

$$\overline{X}_j - \overline{\overline{X}}$$

Sample #1: $45 - 45 = 0$; Sample #2: $35 - 45 = -10$; Sample #3: $55 - 45 = 10$

– Then, square each of these deviations

$$(0)^2 = 0 \qquad (-10)^2 = 100 \qquad (10)^2 = 100$$

– Multiply each squared Deviation by the Group size

$$0 \times 4 = 0 \qquad 100 \times 4 = 400 \qquad 100 \times 4 = 400$$

– Sum these: $SSB = 0 + 400 + 400 = 800$

These numbers are graphed below left and are indicative of a comparatively small Variation <u>between</u> the Groups. Notice that the ranges overlap.

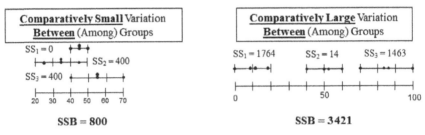

On the right above is a graph of comparatively large Variation <u>between</u> the Groups. There is no overlap in the ranges. We keep saying "comparatively" because, as mentioned earlier, we need to consider both SSW and SSB together in order to come to a definitive conclusion.

If we add SSW and SSB, we get a measure of the total Variation, Sum of Squares Total, SST.

$$\mathbf{SST = SSW + SSB}$$

<u>Notation Alert</u>: some authors use the term "Sum of Squares Treatment" (SST) instead of Sum of Squares Between. That introduces a potential source of confusion, since SST is usually used for Sum of Squares Total.

 3. The Mean Sums (of Squares), MSB and MSW, are averages of SSB and SSW, respectively. **With MSB and MSW, we have only 2 Statistics which summarize the Variation in 3 or more Groups.**

Sums of differences (like SSW and SSB) provide a gross measure of Variation, somewhat analogous to a Range. Averages (Means) are generally more meaningful than sums. (That is why Variances or Standard Deviations are generally more useful than Ranges.) So we calculate the Mean equivalents of SSW and SSB: MSW and MSB.

MSB and MSW are Statistics which each distill information about a type of **Variation involving multiple Groups into a single number**. We can then use these 2 Statistics in a single F-test to accomplish the same thing that multiple *t*-tests would accomplish. Thus, we avoid the compounding of Alpha Error which would occur with multiple *t*-tests.

The downside is that, in calculating the MS's, we lose specific information about the individual Groups. **This is why ANOVA will tell us whether there is a Statistically Significant difference among several Groups, but it will not tell us which one(s) are different.**

> **4. Mean Sums of Squares are similar to the Variance. As such, they can be used to calculate the Test Statistic, *F*,** which is the ratio of two Variances.

Earlier in this article, we said that the generic Sum of Squares is the numerator in the formula for Variance. The denominator in that formula is $n - 1$. As described in the Part 3 article, MSB and MSW are calculated by dividing the Sums of Squares, SSB and SSW, by terms similar to $n - 1$. So MSB and MSW are similar to the Variance.

The Test Statistic *F* is the ratio of two Variances. So, ANOVA is able to use the ratio of MSB and MSW in an *F*-test to determine if there is a Statistical Significant difference among the Means of the Groups.

> **5. Mean Sums of Squares are used in the *F*-tests in ANOVA. A large value of MSB,** compared with MSW, **indicates that the Sample Means are not close to each other.** This makes for a large value for *F*, **which makes it more likely that *F* > *F*-critical.**

SSB $\Rightarrow$ MSB $\quad \Rightarrow \quad \dfrac{\text{MSB}}{\text{MSW}} = F \Rightarrow \quad$ If $F \geq F$-critical, $\Rightarrow$ there <u>is</u> a difference.

SSW $\Rightarrow$ MSW $\qquad\qquad\qquad\qquad\qquad$ If $F < F$-critical, $\Rightarrow$ there is <u>no difference</u>.

- The formulas for MSB and MSW are specific implementations of the generic formula for Variance.
- So, MSB divided by MSW is the ratio of two Variances.
- The Test Statistic *F* is the ratio of two Variances.
- ANOVA uses an *F*-Test (F = MSB/MSW) to come to a conclusion.
- If $F \geq F$-Critical, then we conclude that the Mean(s) of one or more Groups have a Statistically Significant difference from the others.

Related Articles in This Book: *ANOVA – Part 1: What It Does*; *ANOVA – Part 3: 1-Way*; *ANOVA – Part 4: 2-Way*; *Sums of Squares*; *Variation/ Variability/Dispersion/Spread*; *Variance*

ANOVA – PART 3 (OF 4): 1-WAY (AKA SINGLE FACTOR)

Summary of Keys to Understanding

Builds on the content of the ANOVA Part 1 and Part 2 articles.

> 1. In 1-Way ANOVA, we study the effect of one <u>Nominal</u> (named) **Variable, x, on the Dependent** <u>Numerical</u> **Variable, y.**

> 2. **Objective:** Determine whether there is a **Statistically Significant difference among the** <u>Means</u> **of 3 or more groups. Do one or more group Means stand out from the rest?**

x:Script	y: sales in first 100 calls										Mean
A	175	50	225	60	180	170	230	45	90	190	141.5
B	95	150	160	75	120	140	250	70	85	180	132.5
C	80	120	95	225	60	110	160	90	120	140	126.5

> 3. **A 7-step method** (summarized graphically below) **performs the analysis.** Spreadsheets or software will do all this, you just provide the data.

SSB ⇨ MSB

SSW ⇨ MSW ⇨ $\dfrac{\textbf{MSB}}{\textbf{MSW}} = F$ ⇨ If $F ≥ F$-critical, ⇨ there <u>is</u> a difference.

If $F < F$-critical, ⇨ there is <u>no difference</u>.

> 4. **The output includes an ANOVA Table like this:**

ANOVA	Cannot Reject Null Hypothesis because $p > 0.05$ (Means are the same.)					
Source of Variation	SS	df	MS	F	p-Value	F-crit
Between Groups	2686.67	2	1343.33	0.376	0.690	3.354
Within Groups	96567.50	27	3576.57			
Total	99254.17	29				

Explanation

Prerequisite articles: ANOVA Part 1 and ANOVA Part 2.

 | **1. In 1-Way ANOVA, we study the effect of one <u>Nominal</u>** (named) **Variable, x, on the <u>Numerical</u> Variable, y.**

A Nominal (aka Categorical) Variable is one whose values are names.
x is the Independent Variable, also called the Factor. y is the Dependent Variable, since its value depends on the value of x. We might say $y = f(x)$, but in ANOVA (unlike in Regression) we are not interested in determining what the function f is.

Three Examples of Variables in 1-Way ANOVA

Nominal Independent Variable, x	values of the x Variable	Numerical Dependent Variable, y
Script used in call center sales calls	"A", "B", "C"	Sales in dollars
Level of Training	Beginner, Intermediate, Advanced	A worker productivity measurement
School District	names of the 6 school districts	Test scores

ANOVA is frequently used in Designed Experiments. (See the articles on *Design of Experiments*.)

 | **2. <u>Objective:</u>** Determine whether there is a **Statistically Significant difference among the <u>Means</u>** of 3 or more groups. **Do one or more group Means stand out from the rest?**

A Sample of data is taken for each of the values of the x Variable, and the Means of the y measurements for each Sample is calculated.

For example, let's say we're starting up a call center to sell a new product. We hire 30 callers of similar background and divide them into 3 groups of 10. Each group was given a different script to use for their opening sales pitches. We recorded their sales in dollars for the first 100 calls. The x Variable is the name of the script, and the y Variable is the sales amount.

x:script	y: sales in first 100 calls										Mean
A	175	50	225	60	180	170	230	45	90	190	141.5
B	95	150	160	75	120	140	250	70	85	180	132.5
C	80	120	95	225	60	110	160	90	120	140	126.5

There are 3 Samples (groups), A, B, and C. Each has 10 data values, for a total of 30.

Script A appears to give the best results and Script C the worst. But are the differences in the 3 Means Statistically Significant? That's what 1-Way ANOVA can tell us.

 | **3. A 7-step method performs the analysis.** Spreadsheets or software will do all this, you just provide the data.

Before collecting data, select a value for Alpha. Most commonly $\alpha = 0.05$ is selected.

Step 1. Calculate the Sum of Squares (SS) for each Sample.

$$SS = \sum (x_i - \bar{x})^2$$

SS is a measure of <u>Variation within one Sample</u>. In fact, it is the numerator in the formula for Variance.

Step 2. Add all these up for all Samples to get the Sum of Squares Within

$$SSW = SS_1 + SS_2 + \cdots + SS_n$$

SSW is a measure of <u>Variation within all the Samples.</u>

Step 3. Calculate the Overall Mean, $(\bar{\bar{X}})$, of <u>all</u> the data values in <u>all</u> Samples.

Forget which data values go with which Samples, just put them all in one bucket and calculate the Mean.

Step 4: Sum up the differences between each Sample Mean and the Overall Mean to get Sum of Squares <u>Between</u>.

$$SSB = \sum n(\bar{X} - \bar{\bar{X}})^2$$

SSB is a measure of how much the Sample Means differ from the Overall Mean. It also contains information on how much the Sample Means differ from each other.

Step 5: Calculate the Mean Sum of Squares Within (MSW) and Between (MSB).

Sums of differences (like SSW and SSB) provide a gross measure of Variation, somewhat analogous to a Range. But it is often not meaningful to compare sums of different numbers of things. Averages (Means) are generally more meaningful than totals. (That is why Variances or Standard Deviations are generally more useful than Ranges.) So we calculate MSW and MSB.

$$\text{MSW} = \frac{\text{SSW}}{N-k} \quad \text{and} \quad \text{MSB} = \frac{\text{SSB}}{k-1}$$

where N is the overall number of data values in <u>all groups</u>, and k is the number of groups. In our example $N = 30$ and $k = 3$.

SSW and SSB are specific types of the generic Sum of Squares, SS. And the formula for SS is the numerator for the formula for Variance, s^2.

$$s^2 = \frac{\sum(x_i - \bar{x})}{n-1} = \frac{\text{SS}}{N-1}$$

So, if we divide the two special types of Sums of Squares, SSW and SSB, by a Degrees-of-Freedom term (like N–k or k–1), it is easy to see that **MSW and MSB are Variances.**

Step 6: Perform an F-test

The crux of ANOVA is comparing the Variation Within groups to the Variation Between (Among) groups. The best way to do a comparison is to calculate a ratio. **The F-statistic is a ratio of** two Variances, **MSB and MSW.**

$$F = \frac{\text{MSB}}{\text{MSW}}$$

Note that **this is a different concept from the usual F-test comparing Variances of two Samples**. In that case, the Null Hypothesis would be that there is not a Statistically Significant difference between the Variances of two Samples. Although MSB and MSW have formulas like Variances, **MSB and MSW contain information about the differences between the <u>Means</u> of the several groups.** They contain no information about the Variances of the groups.

In the F-Test within ANOVA, **the ANOVA Null Hypothesis is that** there is not a Statistically Significant difference between MSB and MSW – that is, **there is not a Statistically Significant difference among the Means of the several Groups.**

Step 7:

As described in the article on the F-test, Alpha determines the value of F-critical, and the F-statistic (calculated from the Sample data) determines the value of the Probability p. Comparing p to α is identical to comparing F and F-critical

If $F \geq F$**-critical** (equivalently, $p \leq \alpha$), then **there <u>is</u> a Statistically Significant difference between the Means of the groups**. (Reject the ANOVA Null Hypothesis.)

If $F < F$**-critical** ($p > \alpha$), then **there is <u>not</u> Statistically Significant difference between the Means of the groups**. (Accept/Fail to Reject the ANOVA Null Hypothesis.)

The 7–Step ANOVA Process summarized in a concept flow diagram:

SSB $\Rightarrow$ MSB
$\Rightarrow$ $\dfrac{\textbf{MSB}}{\textbf{MSW}} = F$ $\Rightarrow$ If $F \geq F$-critical, $\Rightarrow$ there <u>is</u> a difference.
SSW $\Rightarrow$ MSW
If $F < F$-critical, $\Rightarrow$ there is <u>no difference</u>.

🔑 | **4. The output includes an ANOVA Table like this:**

ANOVA	Cannot Reject Null Hypothesis because $p > 0.05$ (Means are the same.)					
Source of Variation	SS	df	MS	F	p-Value	F-crit
Between Groups	2686.67	2	1343.33	0.376	0.690	3.354
Within Groups	96567.50	27	3576.57			
Total	99254.17	29				

The conclusion of this ANOVA is stated at the top. Prior to the test, Alpha (α) was selected to be 0.05. We see that the p-Value (p) is 0.690, which is greater than Alpha (0.05). So, we do not reject the Null Hypothesis.

Details are given in the table beneath the conclusion about the Null Hypothesis:

"SS" stands for Sum of Squares, and values are given for Between Groups (SSB) and Within Groups (SSW).

"df" is Degrees of Freedom. For Between Groups, df $= k - 1$, where k is the number of groups. In our example, k is 3, so df $= 3 - 1 = 2$. For Within Groups df $= N - k$, where N is the total number (30) of y measurements, so df $= 30 - 3 = 27$.

"MS" is Mean Sum of Squares, and values are given for MSB and MSW. You can see that their ratio gives us F.

$F < F$-critical, which is statistically equivalent to $p > \alpha$.

You might remember that the Part 1 article said that an ANOVA assumption was Continuous, not Discrete data. And the data in this example appear to be Discrete, being in increments of dollars. However, Discrete data in money, which tend to have a large number of possible values, are effectively Continuous.

Related Articles in This Book: *ANOVA – Part 1: What It Does*; *ANOVA – Part 2: How It Does It*; *ANOVA – Part 4: 2-Way*; *Variation/ Variability/Dispersion/Spread*; *Variance*; *F*; *Sums of Squares*; *Critical Values*; *Alpha(α)*; *p-Value*; *ANOVA vs. Regression*; *p, t, and F: ">" or "<"?*

ANOVA – PART 4 (OF 4): 2-WAY (AKA 2-FACTOR)

Summary of Keys to Understanding

Builds on information in the article ANOVA: Part 3 – 1-Way.

1. In **2-Way ANOVA**, we study the effect of **2** <u>Nominal</u> (named) **Variables, A and B,** on the Dependent <u>Numerical</u> **Variable,** *y*.

   A and **B** are **Factors** influencing the value of *y*. "AB" – the **Interaction between and A and B** – can be the 3rd **Factor.**

2. There are **2 Methods** for 2-way ANOVA
   - The <u>WITHOUT Replication</u> method can be used if there is **no Interaction which is Statistically Significant.**
   - Otherwise, you must use the <u>WITH Replication</u> method.

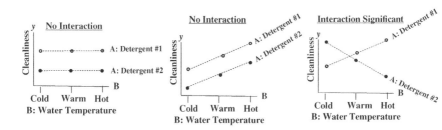

3. **2-Way ANOVA** <u>WITH</u> **Replication simply repeats** (replicates) **the experiment several times** for each combination of A and B values **in order to obtain sufficient data to identify and quantify any Interaction, AB.**

4. In an **ANOVA Table,** $p \leq \alpha$ indicates **Statistical Significance. If the Interaction, AB, is Statistically Significant,** then *p*-values for A and B are not usable.

47

Explanation

> 1. In 2-Way ANOVA, we study the effect of 2 <u>Nominal</u> (named) Variables, A and B, on the Dependent <u>Numerical</u> Variable, *y*.
>
> A and B are Factors influencing the value of *y*. "AB" – the Interaction between and A and B – can be the 3rd Factor

2-Way ANOVA is more complicated – and potentially much more confusing – than 1-Way ANOVA. So, we're going to proceed slowly and deliberately with descriptions of the individual elements involved.

First of all, the names used for different types of Variables can be confusing.

We're familiar with equations of the type

$$y = f(x) \text{ or } y = f(x_1, x_2, \ldots, x_n).$$

The value of the Variable *y* is a function of one or more *x* Variables. In other words, the value of *y* is dependent on the value of one or more *x*'s. So, *y* is called the Dependent Variable. The *x*'s can vary independently and are called Independent Variables.

In 2-Way ANOVA, the equation is of the type

$$y = f(A, B, AB)$$

- *y* **is the Dependent Variable** (also known as the **"Outcome Variable"**).

  <u>*y* is a Numerical Variable.</u> That is, its value is **a Number**, like 5, not a Name, like "Detergent #1."

- **A and B are Nominal (named) Variables.** That is, their values are Names (hence "nominal") within a Category. (Nominal Variables are also known as Categorical Variables.)

  o For example, if the Category A is type of detergent, the values of A would be names or labels for two detergents, say "Detergent #1" and "Detergent #2."

  o B, the second Category could be water temperature. It may have values of "Cold," "Warm," and "Hot." Note, that although these names may have corresponding numerical temperatures (say 40, 80, and 120 degrees Fahrenheit) we do no calculations with those numbers. We are <u>naming</u> 3 <u>levels</u> of temperature, but the numbers behind these names are not used.

- **A, B, and AB are Factors.** We don't use the term Independent Variable, in this context, because AB is not independent of A and B. **A and B** are also called **"Main Effects,"** to distinguish them from Interaction Factors like AB.
- <u>**AB is the Interaction**</u> of A and B. It has an effect on the Outcome Variable different from the effects of A or B separately. As we'll see later, **if the Interaction term is Statistically Significant, then the individual effects of A and B cannot be separately measured.**

<u>Interaction:</u>

Sometimes Factors interact synergistically, that is, the effect of the two of them together is more than just the sum of the effects of each individually. For example, some detergents work much better in hot water than in cold water.

Interacting Factors can also cancel each other out – as in two cleaners, one an acid and the other a base.

2. There are 2 Methods for 2-Way ANOVA
- The <u>WITHOUT Replication</u> method can be used if there is no Interaction which is Statistically Significant.
- Otherwise, you must use the <u>WITH Replication</u> method

In 1-Way ANOVA, we worked with Samples of data in a Population. In 2-Way ANOVA, we **design experiments** to ensure that we get the kind of data that can be analyzed the way we need. For example, we select 2 different detergents and 3 levels of temperature. The numerical Outcome, y, is "Cleanliness," measured on a scale of 0 to 50.

There are 2 methods that can be used for 2-Way ANOVA. **The WITH Replication method is usually better**, because it uses more data and provides more information. However if the experimental budget and time are constraints, **the WITHOUT Replication method can be used, but only if there is no Interaction between the 2 Factors.**

How do we know if there is no Interaction? **Plot the data. If the lines don't intersect there is not a Statistically Significant Interaction.**

Parallel or roughly parallel lines imply no Interaction.
Crossed lines imply Interaction.

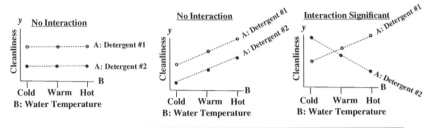

> **Separated** Lines show that **Factor A** has an effect on the Outcome Variable *y*.
> **Slanted** Lines show that **Factor B** has an effect.

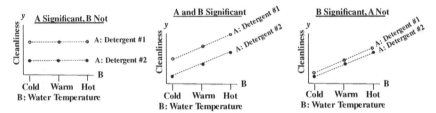

If Factor A has *i* number of values and Factor B has *j*, then there are *i* × *j* pairs of combinations to test. In this example there are 2 values for A: Detergent and 3 values for B: Water Temperature, so there are 2 × 3 = 6 pairs of combinations to test – yielding 6 values of *y* (the numbers in the table above).

The WITHOUT Replication method measures only one value of *y* for each of these combinations. Here is the data we would enter into a spreadsheet or software.

	Cold	Warm	Hot
Detergent #1	30	36	45
Detergent #2	20	29	35

Here is the ANOVA Table produced. (The format and labels will vary somewhat by the tool.)

ANOVA Table: 2-Way WITHOUT Replication (Alpha = 0.05)						
Source of Variation	SS	df	MS	*F*	*p*-value	*F*-crit
Rows (A)	121.5	1	121.5	81	0.012	18.51282
Columns (B)	225	2	112.5	75	0.013	19
Error	3	2	1.5			
Total	349.5	5				

The key items in the ANOVA table are the *p*-values. In the above example, *p*-values for both Rows (Factor A) and Columns (Factor B) are less than 0.05 (the value selected for Alpha), so both have Statistically Significant effects.

Error is the Variation left over after totaling up the Variations caused by A and by B.

Sum of Squares (SS) is the measure of Variation shown. That column shows how much of the total Variation in *y* is caused by Factors A (Rows) and B (Columns), and how much is left over as Error.

Degrees of Freedom (df), and Mean Sums of Squares (MS) are provided for your information. They are used in interim calculations in producing values for the *p*-value and *F*.

And, as is explained in the article, *Critical Values*, $F \geq F$-critical is statistically identical to $p \leq \alpha$. So that information is redundant.

The WITHOUT Replication method has lower experimental costs, but it is limited – **it does not identify or quantify Interactions.**

 3. 2-Way ANOVA <u>WITH</u> Replication simply repeats (replicates) **the experiment several times** for each combination of A and B values **in order to obtain sufficient data to identify and quantify any Interaction, AB.**

Suppose we collected data which produced the graph below.

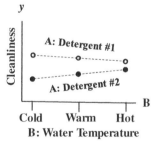

- The lines are separated. But are they separated enough for us to say that Factor B has a Statistically Significant effect?

- The lines are slanted, indicating that Temperature has an effect. But is it a Statistically Significant effect?

- The two lines don't cross, but, if extended, they would. Does this indicate a Statistically Significant Interaction?

The WITHOUT Replication method could answer the first two. But the graph is ambiguous enough that we may want the greater accuracy to be achieved by using more data points, as with the WITH Replication method. That could also answer the question of whether or not there is an Interaction.

The WITH Replication method repeats (Replicates) the experiment several times for each combination of A and B values. That can provide sufficient data to identify and quantify an Interaction. The number of Replications required to achieve a specified level of accuracy is determined by the methods of **Design of Experiments, DOE.** (This book has a 3-part series of articles on DOE.)

Here's the data.

Data for <u>WITH</u> Replication method			
	Factor B		
	Cold	**Warm**	**Hot**
Detergent #1	40, 42, 39	35, 33, 36	30, 29, 31
Detergent #2	20, 18, 21	24, 26, 23	28, 27, 29

(Factor A)

4. **In an ANOVA Table, $p \leq \alpha$ indicates Statistical Significance. If – in the WITH Replication method – the Interaction, AB, is Statistically Significant, then p-values for A and B are not usable.**

Here's the ANOVA Table, which is calculated from WITH Replication data above:

ANOVA Table: 2-Way With Replication (Alpha = 0.05)							
Source of Variation	SS	df	MS	F	p-value	F-crit	
Sample (A)	93.4	1	93.4	5.4	0.038	4.7	Reject Null Hypothesis
Columns (B)	310.3	2	155.2	9.0	0.004	3.9	Reject Null Hypothesis
Interaction (AB)	14.8	2	7.4	0.4	0.660	3.9	Do Not Reject Null Hypothesis
Within	206.0	12	17.2				
Total	624.5	17					

The first thing we do is check the p-value for the Interaction AB.

If that p-value indicates a Statistically Significant Interaction ($p \leq$ Alpha), then the p-values calculated for A and B separately would be meaningless. The effects of A and B would be too intertwined to separate them

In this example, the $p > \alpha$, so we do not reject the Null Hypothesis. The Null Hypothesis of No Statistically Significant effect for the Interaction is supported by the analysis.

(Note that there are 3 different Null Hypotheses here: one each for Factor A, Factor B, and the Interaction AB.)

Since there is no Statistically Significant Interaction, **we can check the p-values for the two Factors, A and B. If $p \leq \alpha$, then that Factor does have a Statistically Significant Effect.** The Null Hypothesis of No Statistically Significant effect for the Factor is Rejected.

In this example, the p-values for both Factors A and B are less than Alpha. So, we Reject the Null Hypothesis and conclude that both the Factors A and B have a Statistically Significant effect on the outcome Variable y.

Related Articles in This Book: *ANOVA: Part 1 – What It Is*; *ANOVA: Part 2 – What It Does*; *ANOVA: Part 3 – 1-Way*; *ANOVA vs. Regression*; *Design of Experiments – Parts 1–3*; *F*; *Variation/Variability/Dispersion/ Spread*; *p, p-Value*; *Alpha (α)*; *Critical Value*

ANOVA vs. REGRESSION

The purpose of this article is to give you a more intuitive understanding of both ANOVA and Regression by exploring how they are similar and how they differ.

Summary of Keys to Understanding

	ANOVA	Regression
1. Purpose	Determine whether the Means of 2 or more Populations are statistically the same.	Model Cause and Effect; Predict y value from x value(s).
2. Type of Question	Is there a Statistically Significant difference between drugs A, B, and placebo?	How much do house prices increase as the number of bedrooms increases?
3. Variable Types	x: Categorical, y: Numerical	x and y both Numerical
4. Groups Being Compared	Individual Populations (or Samples of each)	data values for the y Variable vs. corresponding y values on the Regression Line
5. Focuses on Variation	Yes	Yes
6. Uses Sums of Squares to Partition Variation	Yes	Yes
7. Variation of …	Means of Different Populations	Dependent Variable (y) vs. Independent Variable(s) (x's)
8. Involves Correlation	No	Yes
9. Sum of Squares Total (SST) =	SSW + SSB	SSR + SSE
10. Key Sum of Squares Ratio	F = MSB/MSW	R^2 = SSR/SST
11. Analysis Output Includes ANOVA Table	Yes	Yes
12. Used Primarily In	Designed Experiments	Inferential Statistics, but validated via Designed Experiments

Explanation

ANOVA and Regression have a number of similarities and differences. The purpose of this article is to give you a more intuitive understanding of both ANOVA and Regression by exploring both how they overlap and how they differ. Let's start with some key differences.

	ANOVA	**Regression**
1. Purpose	Determine whether the Means of 2 or more Populations are statistically the same.	Model Cause and Effect; Predict y value from x value(s).
2. Type of Question	Is there a Statistically Significant difference between Drug A, Drug B, and Placebo?	How much do house prices increase as the number of bedrooms increase?

ANOVA and Regression differ in their purposes and in the type of question they answer.

ANOVA:
ANOVA is actually more similar to the t-test than to Regression. ANOVA and the 2-Sample t-test do the same thing if there are only 2 Populations – they **determine whether the Means** of the 2 Populations **are statistically the same or different**.

This, then, becomes **a way of determining whether the 2 Populations are the same or different – relative to the question being asked.** ANOVA can also answer the question for 3 or more Populations.

The answer to the question is Yes or No.

Regression:
The purpose of Regression is very different. It attempts to produce a Model (an equation for a Regression Line or Curve) which can be used to **predict the values of the y (Dependent) Variable given values of one or more x (Independent) Variables.**

Regression goes beyond mere Correlation (which does not imply Causation) to attempt to **establish a Cause and Effect relationship** between the x Variable(s) and the values of y.

   The answer to the question is the equation, for the best-fit Regression Line, e.g., House Price = $200,000 + ($50,000 × Number of Bedrooms).

	ANOVA	Regression
3. Variable Types	x: Categorical, y: Numerical	x and y both Numerical

ANOVA
The Independent Variables (x) Must be Categorical (Nominal). That is, the different values of x in the category (e.g., drug) must be names (e.g., Drug A, Drug B, Drug C, Placebo), rather than numbers.

   The Dependent Variable (y) must be Numerical, e.g., a blood pressure measurement.

Regression
Both the Independent and Dependent Variables must be Numerical. For example, x is Number of Bathrooms and y is House Price. As mentioned earlier, Regression attempts to establish a Cause and Effect relationship, that is, increasing the number of Bathrooms results in an increase in House Price.

	ANOVA	Regression
4. Groups Being Compared	Individual Populations (or Samples of each)	data values for the y Variable vs. corresponding y values on the Regression Line

   Regression really doesn't compare groups as such. But if one wants to explore this similarity between Regression and ANOVA, one would describe Regression concepts in terms used by ANOVA.

   We can consider the Sample of paired (x, y) data to represent one group. And the other group consists of corresponding paired (x, y) points on the Regression Line. By "corresponding" we mean having the same x values.

   <u>Illustration:</u> 7 pairs of (x, y) data and their corresponding points on the Regression Line.

   The Regression Line is $y = 2x$. We take the value of x from a data point, and calculate the y value for the Regression Population using $y = 2x$

Group 1	Data Points (x, y)	(1, 2.5)	(2, 1.9)	(4.7)	(5, 9)	(7, 15)	(8, 18)	(11, 22)
Group 2	Corresponding Regression Points	(1, 2)	(2, 2)	(4, 8)	(5, 10)	(7, 14)	(8, 16)	(11, 20)

ANOVA will compare the Means of the y values of these groups.

	ANOVA	**Regression**
5. Focuses on Variation	Yes	Yes
6. Uses Sums of Squares to Partition Variation	Yes	Yes

The main conceptual similarity between **ANOVA and Regression** is that they **both analyze Variation to come to their conclusions.**

"Partitioning" Variation Means dividing up the Total Variation – as measured by **Sum of Squares Total (SST)** – into components or portions of the total Variation.

	ANOVA	**Regression**
7. Variation of …	Means of Different Populations	Dependent Variable (y) vs. Independent Variable(s) (x's)
8. Involves Correlation	No	Yes

Both ANOVA and Regression use Variation as a tool. For Regression, we know that the Variables x and y vary – that is, all their values in a Sample will not be identical. That is, a Sample will not be something like (2, 3); (2, 3); (2, 3); (2, 3); (2, 3); (2, 3); (2, 3). **The first question for Regression is, do x and y vary together** – either increasing together, or moving in opposite directions. That is, **is there a Correlation between the x and y Variables?** If there is not a Correlation, then we will not even consider doing a Regression analysis.

For ANOVA, there is no question of "varying together," because the values of the x Variable – being a Categorical Variable. They don't increase or decrease.

	ANOVA	**Regression**
9. SST =	SSW + SSB	SSR + SSE

Since ANOVA and Regression measure very different types of Variation, one would expect that the components of their total Variations are very different.

ANOVA: SST = SSW + SSB where SST is **Sum of Squares Total,** SSW is **Sum of Squares Within,** and SSB is **Sum of Squares Between**

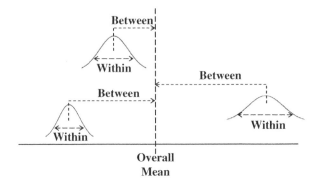

The total Variation (SST) is the sum of all the Variations **within** each of the individual Populations plus the sum of the Variations **between** each Population Mean and the Overall Mean.

Regression:

For Regression, the two components of SST are Sum of Squares Error (SSE) and Sum of Squares Regression (SSR). We use the data to calulate one component, SSE, and to calculate the total, SST. Then, we calculate the other component, SSR from SST and SSE:

In this very simple example, there are only 3 data points in our sample. These are illustrated by the 3 <u>black</u> <u>dots</u>. The 3 data points have x, y values of (2,6), (1,2), and (0,1). The Regression line is defined by formula $y = 3x$.

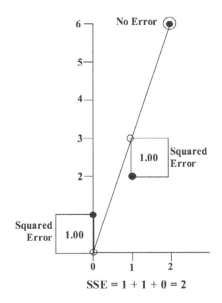

There is <u>no error</u> for the point at the top 2, 6. It is <u>on</u> the Regression line of y = 3x.

The black dots of the other two points, (1,2) and (0,1) are each one unit away from the Regression line. So, their error is 1 and their squared error is also 1.

And the Sum of these Squared Errors, SSE, is 0 + 1 + 1, which equals 2.

Now, let's look at Sum of Squares Total, SST.

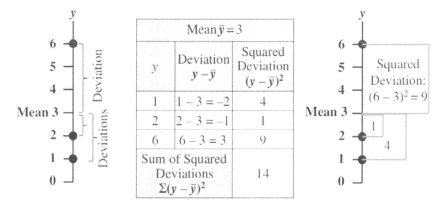

SST, is the sum of the squared deviations of the data values of the Variable y to the Mean of y.

As shown as black dots in the vertical graph on the left, our 3 data points had y values of 1, 2, and 6.

They are also shown in the first column of the table in the middle.

1 + 2 + 6 = 9, divided by 3 gives us a Mean value of 3 for the y Variable, as stated in the top row of the table.

The middle column of the table calculates the 3 deviations from this Mean, -2, -1 and 3.

And the right column of the table shows the squared deviations, 4, 1, and 9. This is also illustrated in the diagram to the right of the table.

The sum of the Squared deviations is $4 + 1 + 9 = 14$. This is SST, the Sum of Squares Total. Given SST and SSE, we can calculate SSR, the Sum of Squares Regression.

$$SSR = SST - SSE = 14 - 2 = 12$$

	ANOVA	Regression
10. Key Sum of Squares Ratio	$F = MSB/MSW$	$R^2 = SSR/SST$

ANOVA: F = MSB/MSW,

where MSW is the **Mean Sum of Squares Within** and MSB is the **Mean Sum of Squares Between.** These are calculated by dividing SSW and SSB, respectively, by their Degrees of Freedom. **MSW and MSB are different types of the Statistic, Variance.**

The F-statistic is a ratio of two Variances – MSW and MSB, in this case. Comparing F to its Critical Value tells us whether there is a Statistically Significant difference among the (Means of) the Groups being compared.

SSB ⟹ MSB   ⟹ $\dfrac{\textbf{MSB}}{\textbf{MSW}} = F$ ⟹   If $F \geq F$-critical, ⟹ there <u>is</u> a difference.

SSW ⟹ MSW   If $F < F$-critical, ⟹ there is <u>no difference</u>.

Regression: R^2 = SSR/SST,

where **SSR is the Sum of Squares Regression**. SSR is the component of the Variation in the **Total Variation in the y Variable (SST)** which is explained by the Regression Line. SSR/SST is the proportion.

R^2 **is a measure of the Goodness of Fit of the Regression Line.** If R^2 is greater than a predetermined clip level, then the Regression Model is considered good enough, and its predictions can then be subjected to validation via Designed Experiments.

	ANOVA	Regression
11. Analysis Output Includes ANOVA Table	Yes	Yes

Spreadsheets and statistical software often include an ANOVA table in their outputs for both ANOVA and for Regression:

ANOVA Table from a Regression analysis					
	df	SS	MS	F	p-value
Regression	2	48,845.938	24,422.969	64.040	0.001
Residual	4	1525.490	381.373		
Total	6	50,371.429			

SS for Regression is SSR, SS for Residual is SSE ("Residual" is another name for Error) and SS for Total is SST.

Divide the SS's by the df's (Degrees of Freedom) to get the MS's (Mean Sums of Squares for Regression and Error). The F-statistic is MS Regression/MS Residual.

This particular table doesn't show the Critical Value of F with which to compare the value of F. But it does show the p-value, which can be compared to the value we selected for the Significance Level, Alpha (α).

So, in this example **p is much less than Alpha, so we can conclude that** the results are Statistically Significant. That's another way of saying **the Regression Line is a good fit for the data**. This was confirmed by the value (not shown in the ANOVA table) of $R^2 = 0.893$

	ANOVA	**Regression**
12. Used Primarily In	Designed Experiments	Inferential Statistics, but validated in Designed Experiments

One of the most significant differences between ANOVA and Regression is in how they are used. ANOVA has a wide variety of uses. It is well-suited for Designed Experiments, in which levels of the x Variable can be controlled – for example, testing the effects of specific dosages of drugs.

Regression can be used to draw conclusions about a Population, based on Sample data (Inferential Statistics). The purpose of Regression is to provide a Cause and Effect Model – an equation for a Best Fit Regression line or curve – which predicts a value for the y Variable from a value of the x Variable(s). Subsequent to that, data can be collected in Designed Experiments to prove or disprove the validity of the Model.

Related Articles in This Book: *ANOVA – Parts 1–4*; *Regression – Parts 1–5*; *r, Multiple R, R^2, R Square, Adjusted R^2*; *Sum of Squares*

BINOMIAL DISTRIBUTION

Summary of Keys to Understanding

 1. The Binomial Distribution is used with **Discrete data**. It displays the **Probabilities of Counts of outcomes** of Binomial Experiments. **Units are counted**, not Occurrences.

 2. In a Binomial Experiment,
   a. There are **a fixed number, _n_, of trials.**
   b. Each trial can have only one of two outcomes – Yes or No.
   c. **The Probability, _p_,** of a Yes in each trial **is the same for all trials.**
   d. **Each trial is Independent** of the others. This means the sampling is done **With Replacement.**

 3. There are **different Binomial Distributions for different values of _n_** (the number of trials) **and _p_** (the Probability of each trial).

 4. **Mean: $\mu = np$;   Standard Deviation: $\sigma = \sqrt{np(1-p)}$**

 5. **The Binomial Distribution is useful for solving problems of the kind:**
   What is the Probability that a Sample of 10 units will include 2 or more defective units?

 6. **Under specific conditions, the Binomial can be related to the Hypergeometric, Poisson, or Normal Distributions.**

Statistics from A to Z: Confusing Concepts Clarified, First Edition. Andrew A. Jawlik.
© 2016 John Wiley & Sons, Inc. Published 2016 by John Wiley & Sons, Inc.

Explanation

 > **1.** The Binomial Distribution is used with **Discrete data**. It displays the **Probabilities of Counts of outcomes** of Binomial Experiments. **Units are counted**, not Occurrences.

Discrete data are integers, such as Counts. Counts are non-negative. In contrast to Continuous data, there are no intermediate values between consecutive integer values.

The Binomial Distribution is used for Counts of Units, such as the number of shirts manufactured with defects. Units are different from Occurrences. If a shirt (the Unit) we inspected had 3 defects, we would add only 1 to the Count of defective Units, and we could use the Binomial Distribution.

(If we were interested in the total number of Occurrences of defects – not the number of defective Units – we would count 3 Occurrences for that shirt and we would use a different Discrete data Distribution – the Poisson Distribution.)

 > **2.** In a Binomial Experiment,
> a. There are **a fixed number, n, of trials**.
> b. Each trial can have only one of two outcomes – Yes or No.
> c. **The Probability, p,** of a Yes in each trial **is the same for all trials**.
> d. **Each trial is Independent** of the others. This Means the sampling is done **With Replacement.**

a. An example of a trial would be a single coin flip. In a Binomial Experiment there is **a fixed number, n, of trials.**

b. Each trial can have only one of two outcomes. In this book, we will call them "Yes" and "No." **In the trials, we will count only one of the two outcomes**. In a series of coin flip trials, we may choose to count the number of heads. So we pose the question: is the coin flip a head? If Yes, we add one to the Count.

Terminology:

– Note this "p" is a different concept from the "p" aka "p-value" which is compared to Alpha in Inferential Statistics. They are both Probabilities, but p-value is the Probability of an Alpha Error.

– Pretty much every other book or web page you might read would call the outcomes "Success" or "Failure." This results in the bizarre practice of calling a quality failure a "Success" when a trial outcome observes a defect. To avoid this confusion, we'll say "Yes" instead of "Success" and "No" instead of "Failure."

– Another name for Binomial Experiment is "Bernoulli Experiment."

c. Another requirement is that **the Probability for each trial is the same** – as in a coin flip. Each time you flip a coin, the Probability of a head is 50%, 0.50. This doesn't change.

d. Also, the outcomes of previous coin flips have no influence or those that follow. That is, **each trial is Independent of the others**. If you get 10 tails in a row, the Probability of a Yes (head) in the next coin flip is still 50%. So, each trial is Independent.

The concept of Independent trials is related to the concept of Sampling with Replacement.

To illustrate this, let's say we're doing a study in a small lake to determine the Proportion of Lake Trout. Each trial consists of catching and identifying one fish. If it's a Lake Trout, we count one Yes. The Population of the fish in the lake is finite. We have no way of knowing from our boat on the lake, but let's say there happen to be 100 fish in the lake, 70 Lake Trout, and 30 Rainbow Trout.

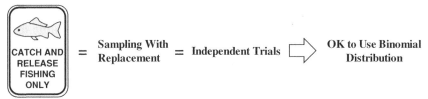

If we throw the fish back in before trying to catch the next fish, that is called **Sampling <u>With</u> Replacement.** Each time we drop our line to catch a fish (a trial), the Probability of catching a Lake Trout remains 70%. The Probability of each trial does <u>not</u> depend on the outcomes of any other trials. **The trials are Independent of each other. That is a requirement for a Binomial Experiment.**

But, if we keep the fish – Sample <u>without</u> Replacement – the situation changes. Let's say that the first 5 fish which we catch (and keep) are Lake Trout. There are now 95 fish in the lake – 65 Lake Trout and 30 Rainbow Trout. The Probability of a Yes in the next trial is 65/95 = 68.4%; this is a change from the original 70%.

So, **Sampling <u>Without</u> Replacement causes the trials to <u>not</u> be Independent**, and we do <u>not</u> have a Binomial Experiment. **We cannot use the Binomial Distribution for Sampling Without Replacement; we must**

use the Hypergeometric Distribution instead. (See the article *Hypergeometric Distribution.*)

   Sampling a Population or Process that is "infinite" is Sampling With Replacement. An ongoing Process, like continuing to flip coins, can be considered infinite.

> **3.** There are **different Binomial Distributions for different values of** n (the number of trials) **and** p (the Probability of each trial).

There are infinitely many Distributions in the family of Binomial Distributions – one for each combination of the values of n (the number of trials) and p (the Probability) of each trial. In the graphs that follow, the horizontal axis is the Count of Units, denoted by X. The vertical axis is the Probability of that Count, denoted by $\Pr(X)$.

   The three graphs below show the effect that p has on the Binomial Distribution. For $p = 0.5$ (50%), the left graph shows that the Distribution is symmetrical about the Mean. For $p < 0.5$, the mass of the Distribution is on the left, and the tail is skewed to the right. $p > 0.5$ has the opposite effect.

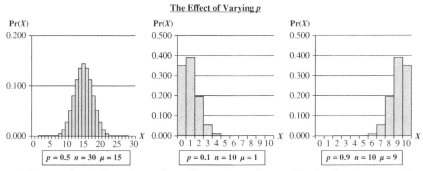

   The graphs below show the effect of varying n. Basically, it moves the Distribution to the right.

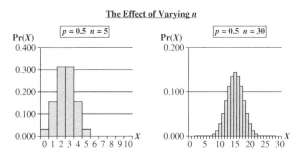

So, the Probability of a given value of X is a function of X, p, and n.

$$\text{Pr}(X) = f(X, p, n)$$

(Note that we use a capitalized "X" for Discrete Variables, while a lower case "x" is used for Continuous Variables.) Here's what that function looks like:

$$\text{Pr}(X) = \frac{n!}{X!(n-X)!} p^X (1-p)^{n-X}$$

This is the formula for the Probability of a Count of exactly X *Yesses* in n trials.

Other books and web pages can take you through its derivation. It's a multiple-step procedure that doesn't really add to an intuitive understanding for most people. Besides, you never have to use it, because spreadsheets and software are available to do it for you.

Distributions can often be succinctly described by their Mean and Standard Deviation. In contrast to the Probability formula above, the formulas for the Mean and Standard Deviation can help us get a more intuitive understanding. So let's take a look at them.

 | **4. Mean: $\mu = np$;   Standard Deviation: $\sigma = \sqrt{np(1-p)}$** |

Without knowing anything about statistics, if you were to flip a coin 30 times, and you had to guess how many heads there would be, what would you guess? Almost everybody would say 15. Intuitively, you calculated np. You know that it may not be exactly 15, but it would most probably be close to 15. On average you would expect it to be 15. So, it makes intuitive sense that the average, or Mean, of a Binomial Distribution would be *np*.

As we can see in the left graph below, the Mean is 15 and values close to 15 also have high Probabilities. The middle and right graphs illustrate $np = 1$ and $np = 9$ as the Means.

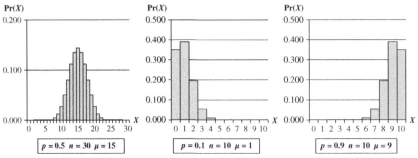

The Standard Deviation is a measure of Variation about the Mean — how spread-out a Distribution is from the Mean. The formula for the Standard

Deviation of a Binomial Distribution is

$$\sigma = \sqrt{np(1 - p)}$$

The first thing this formula tells us is that the **Standard Deviation grows larger as *n* grows larger.** This makes sense, because if $n = 5$, then the possible Counts, the X's, can range only from 0 to 5. If $n = 30$, Counts can range from 0 to 30.

The effect of $p(1 - p)$ is less intuitive. But we can see from the table below that the largest values for $p(1 - p)$ are produced when p is closest to 0.5.

p	0.1	0.2	0.3	0.4	0.5	0.6	0.7	0.8	0.9
$1 - p$	0.9	0.8	0.7	0.6	0.5	0.4	0.3	0.2	0.1
$p(1 - p)$	0.09	0.16	0.21	0.24	0.25	0.24	0.21	0.16	0.09

So, **the Standard Deviation of a Binomial Distribution gets larger as *p* gets closer to 0.5.**

> **5. The Binomial Distribution is useful for solving problems of the kind:**
>
> What is the Probability that a Sample of 10 Units will include 2 or more defective Units?

Let's say a Process has historically produced products with a defective unit rate of $p = 0.02$. We take a Sample of $n = 10$ units, and we find 1 unit is defective. We're wondering if something has happened to the Process. So, we want to know: what is the Probability of 1 or more defective units?

This is the Probability of $X = 1$ defective unit + the Probability of 2 defective units + ... + the Probability of 10 defective units. More simply, it is 1 − the Probability of 0 units.

From a table or software we find that $Pr(0) = 0.817$. So, the Probability of getting 1 or more defective units is $1 - 0.817 = 0.183$. So we can expect a Sample of 10 to have 1 or more defects about 18% of the time. This is not strong evidence of a change in the defective unit rate in the Process – if we're used to thinking in terms of a Probability less than the Level of Significance of 5% (0.05).

But what if we found 2 defective units? The $Pr(1) = 0.167$. So, the Probability of 2 or more defective units is $1 - [Pr(0) + Pr(1)] = 1 - [0.817 + 0.167] = 0.016$, which is less than 2%. So, 2 defective units in a Sample of 10 would be strong evidence that we have a problem with the Process.

> **6. Under specific conditions, the Binomial can be related to the Hypergeometric, Poisson, or Normal Distributions.**

We said earlier that when sampling **Without Replacement,** we should use the Hypergeometric Distributions instead of the Binomial. However, **the Binomial can be used as an approximation for the Hypergeometric when the Population Size (N) is large relative to the Sample Size (n), for example, when $N > 10n$.** This makes sense, because Replacement of a Sample back into the Population has a small impact when the Sample is very small and the Population is very large.

The Binomial Distribution approaches the Poisson Distribution as n approaches infinity, while p approaches zero, (keeping np fixed). The Poisson Distribution is another Discrete data Distribution; it is used when counting Occurrences, not Units.

Also, **the Binomial approaches the Normal Distribution as n approaches infinity** (while keeping p fixed).

Related Articles in This Book: *Distributions – Parts 1–3*; *Hypergeometric Distribution*; *Normal Distribution*; *Poisson Distribution*; *Standard Deviation*

CHARTS/GRAPHS/PLOTS – WHICH TO USE WHEN

For this:	Use this:
1 Variable: Shape of the data Distribution	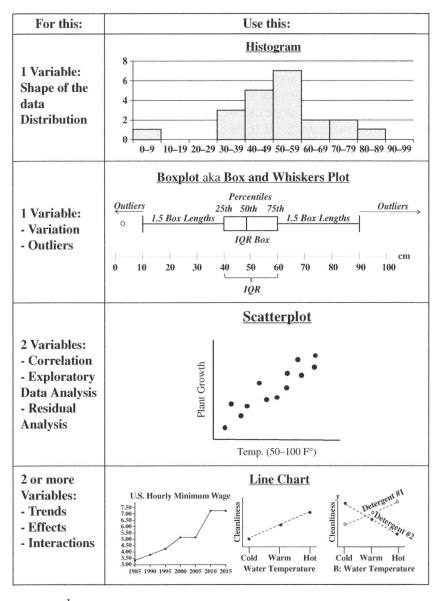
1 Variable: - Variation - Outliers	
2 Variables: - Correlation - Exploratory Data Analysis - Residual Analysis	
2 or more Variables: - Trends - Effects - Interactions	

... and more

Statistics from A to Z: Confusing Concepts Clarified, First Edition. Andrew A. Jawlik.
© 2016 John Wiley & Sons, Inc. Published 2016 by John Wiley & Sons, Inc.

Explanation

It has been said that the first three rules of Statistics are: #1. Plot the data #2. Plot the data, and #3: Plot the data. (Alternately: #1 Draw a picture, #2 Draw a picture, #3 Draw a picture – of the data.)

"Chart," "graph," and "plot" are three words for the pictures we can make from data. They can make patterns (or lack thereof) apparent that just analyzing the numbers would not uncover. So, they are extremely useful in getting an insight into what the data mean.

Calculated statistics alone can be misleading. For example, we plotted the following two data sets in the article *Correlation – Part 2*.

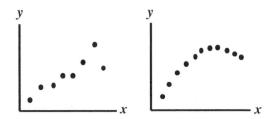

The first graph would indicate that there is a roughly linear Correlation between the x and y Variables. (As the value of x increases, the corresponding value of y increases – roughly following a diagonal straight line.) The second plot shows data that are obviously <u>not</u> linearly correlated. And yet, the calculated Correlation Coefficients for both data sets are almost identical. They both indicate a strong linear Correlation. In deciding how to interpret the data in this case, **the visual interpretation of the graphs takes precedence over the calculated value of the Statistic.**

There are many different kinds of statistical charts. The following are some of the most commonly used. Spreadsheets or statistical software can produce these from your data.

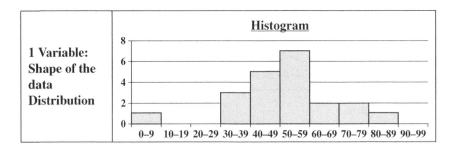

If you have data for a single Variable, call it "x", and you want to get a picture of how its values are distributed, you can use a Histogram.

A Histogram is a bar chart with no spaces between the bars. Each bar represents a range in the values of the Variable. These ranges are also called "bins." We decide limits of the range which define the bins.

In the example above, the ranges are 0 – 9, 10 – 19, etc. If a data value is within the range, 1 is added to the Count represented by the height of the bin. In the example, we can see that there is 1 value between 0 and 9, but no values between 10 and 29. There are 7 values between 50 and 59.

The height of the bars can indicate a Count or a Percentage or a Probability.

A Histogram can give you an indication of

- **Shape**: Is it roughly left–right symmetrical, like a Normal Distribution and that shown in the Histogram above? Or is the long tail Skewed to the right or left? Does it have one hump (Mode) or two or more?

- **Central Tendency**: Where is the Mean, Mode, or Median?

A Histogram is <u>not</u> good for picturing Variation (aka **Spread**). The arbitrary choice of bin range can affect the visual depiction of Spread. We could also squeeze or stretch out the image to make the Spread appear to be smaller or larger.

Dot Plot

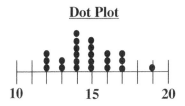

A **Dot Plot** can be used in place of a Histogram for small data sets. It shows each data value as a point on the plot, so no information is lost due to binning.

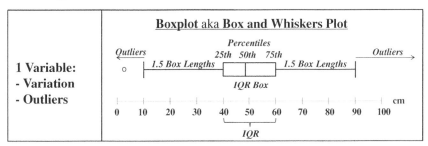

A **Boxplot is a very good way to get a picture of Variation**. In the example above, the IQR box represents the InterQuartile Range, which is a useful measure of Variation. (See the article *Variation/Variability/Dispersion/Spread*.) This plot shows us that 50% of the data points (those between the 25th and 75th Percentiles) were within the range of 40–60 cm.

25% were below 40 and 25% were above 60. The Median, denoted by the vertical line in the box is about 48 cm.

Any data point outside 1.5 box lengths from the box is called an Outlier. Here, the Outlier with a value of 2 cm is shown by a circle. Not shown above, but some plots define an Extreme Outlier as one **that is more than three box lengths outside the box. Those can be shown by an asterisk.**

Showing several vertically oriented Boxplots together is a good way to compare Variation for multiple data sets. In the graph below, we can see that the Medians (the lines in the middle of the IQR boxes) are fairly close for treatments A, B, and C. Treatment A had the highest top–end results. However, both the Box and the Whiskers for Treatment A are quite spread out, indicating a comparatively large amount of Variability – a lack of consistency. Treatment B, on the other hand, has much less Variability. Plus, its lowest whisker is at the 25th percentile of its nearest competitor, Treatment C. So, even without further analysis or study, one could use a set of Boxplots like this to get a strong indication of which is the best treatment.

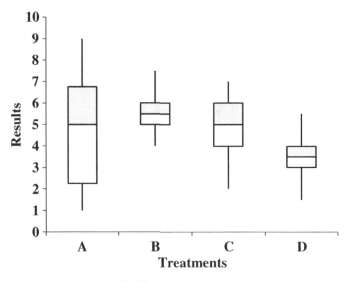

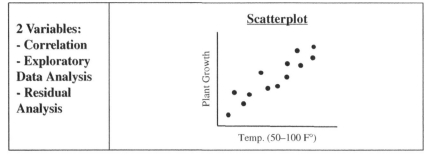

The charts shown so far have plotted Counts of values of a single Variable. Now, we'll cover charts for two Variables. The idea is to see whether and how they interact with each other. Each data point is described by a <u>pair</u> of values representing the values of the two Variables. The data points are plotted in two dimensions. The first value in the pair is usually denoted by *x*, and it is represented along the horizontal axis. The *y* value is represented along the vertical axis.

Correlation

The Scatterplot simply plots the *x*, *y* points. It does not attempt to connect them in any way. However, our minds will often do so. For example, it is easy for us to mentally overlay a diagonal line through the data points above. We infer that there is a Correlation between the two Variables' temperature and plant growth. This is the first step in Correlation analysis. If we do not see a visual Correlation, then we do not proceed to the next step, which is to calculate a Statistic (the Correlation Coefficient) to tell us the strength of the Correlation

Exploratory Data Analysis (EDA)

The Scatterplot is often one of the first steps in EDA. We're looking for insights from data that we can follow up with further statistical analysis or controlled experiments. Is there a potential cause/effect relationship between the variables?

Residual Analysis

If a Regression Model is good, the Residuals (differences between individual points in the calculated model and the corresponding results from a subsequent test) should be randomly distributed. So, a Scatterplot of the Residuals vs. *y* should show no patterns. Likewise, a Scatterplot of Residuals vs. time should show no patterns. (See the article *Residuals*.)

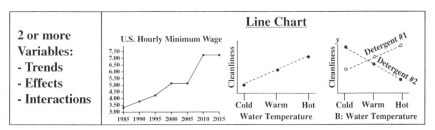

A Line Chart is like a Scatterplot with lines that connect points that have adjacent *x* values. It works best when there are a small number of

data points to be connected. **It is often used to illustrate trends**, with the horizontal axis representing time.

It is also used to graph cause-and-effect, in which the *x* Variable is the Factor which causes the effect in the *y* Variable. In the center chart above, an increase in the Factor Variable, water temperature, causes an increase in the Effect Variable, cleanliness. This is used **in Regression analysis and in the Designed Experiments which are conducted to test a Regression Model.**

The rightmost chart combines two line charts into one. It has the same *x* and *y* Variables as the center chart, but it adds a second Factor (*x*) Variable, Detergent type. So, there are two lines, connecting two sets of data points. In **2-Way ANOVA**, crossing lines indicate that there is an Interaction between the two Factors. In this case, an increase in temperature has the opposite effect for the two detergent types – it makes Detergent #1 do better, and it makes Detergent #2 do worse. If the lines were parallel or did not cross, then there would be no Interaction.

In a similar fashion, a Line Chart can help differentiate between Observed and Expected Frequencies in a **Chi-Square test for Goodness of Fit**.

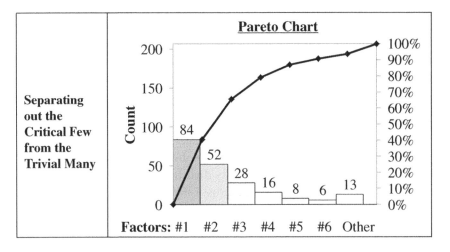

The Pareto Chart helps illustrate the so-called 80/20 "rule": About 80% of the effect is often due to about 20% of the causes. (This is folk wisdom, not a law of statistics.) Among other uses for the Pareto Chart, Multiple Linear Regression involves casting a wide net to identify all possible Factors, and then selecting a few to analyze further.

The Pareto Chart combines two charts – a sorted Bar Chart and a Line Chart. The bars are sorted left to right by Count of Occurrences of the

Effect due to that Factor. The Counts over each bar line up with the "Count" vertical axis on the left. After Factors #5 or 6 or so, it's usually a good idea to lump all the rest under "other."

The line graph shows the cumulative percentage of the total Effect Count which is comprised by the bar below it and the bars to the left. The dot at the right top of each bar lines up with a cumulative percentage on the right axis. In the example, the first three bars comprise about 80% of the total. If you want to get to 90%, you'd have to address twice as many causes.

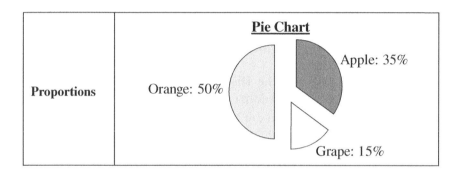

Proportions represent shares of 100% (the whole pie). So, Pie Charts are a good way to depict the relative size of the individual "slices," as shown in this chart of fruit juice preferences.

Related Articles in This Book: *Correlation – Part 2*; *Distributions – Part 1: What They Are*; *Variation/Variability/Dispersion/Spread*; *Residuals*; *Regression – Part 4: Multiple Linear*; *Proportion*

CHI-SQUARE – THE TEST STATISTIC AND ITS DISTRIBUTIONS

The three Chi-Square Tests are covered in separate articles.

Summary of Keys to Understanding

 1. **Chi-Square, χ^2, is a Test Statistic** which is **very versatile in the types of data it can handle:** Discrete, Continuous, non-Normal, Categorical.

 2. As with the F and the t Test Statistics, **there is a different Chi-Square Distribution for each value of Degrees of Freedom** (df).
   - In each case, **the Distribution's Mean is equal to the Degrees of Freedom ($\mu = $ df).**
   - **For larger values of Degrees of Freedom:**
     o **the Distributions move to the right**
     o **they become more symmetrical**
     o **Critical Values increase** (move to the right)
     o **the Variances increase** (the Spread becomes wider).

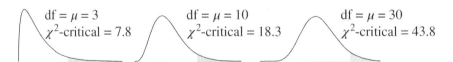

$$df = \mu = 3$$
$$\chi^2\text{-critical} = 7.8$$

$$df = \mu = 10$$
$$\chi^2\text{-critical} = 18.3$$

$$df = \mu = 30$$
$$\chi^2\text{-critical} = 43.8$$

 3. Furthermore, **for All Chi-Square Distributions:**
   - the **Mode = df – 2** (for df $\geq$ 3)
   - the **Variance = 2df; Range: $\chi^2 = 0$ to Infinity**
   - they **approach, but never touch the horizontal axis** as they extend to the right
   - they **are not symmetrical** – they are skewed toward the right tail.

 4. Since Chi-Square Distributions are not symmetrical, **there are two different Critical Values for a 2-sided Chi-Square test.**

 5. **Chi-Square is used in Inferential Statistics to analyze Variances via three different Chi-Square Tests: for the Variance, for Independence, and for Goodness of Fit.**

Explanation

 1. **Chi-Square, χ^2, is a Test Statistic** which is **very versatile in the types of data it can handle:** Discrete, Continuous, non-Normal, Categorical.

Chi-Square, sometimes called "Chi-Squared," is a Test Statistic like z, t, and F. A Test Statistic is one which has a Distribution or Distributions with known Probabilities for every value of the Test Statistic. So, for any value of χ^2 (on the horizontal axis in the diagram below), there is a known Probability of that value occurring (and vice versa). That Probability is the height of the curve above that point.

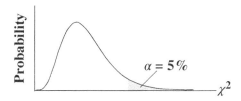

More importantly, we can calculate the area under the curve to the left or right of any value of a Test Statistic. This gives us a Cumulative Probability (such as α or p) which we can use in various types of Inferential Statistical tests involving Hypothesis Testing or Confidence Intervals.

z, t, and F are fairly restrictive about the types of data they can handle. But **Chi-Square** is much more versatile. It **can handle**:

– **Discrete data** (such as Counts of Nominal/Categorical variables. For example, Counts by gender or political party affiliation)
– **Continuous/Measurement data** (e.g., temperature, weight)
– **Non-Normal data**
– **Data in 2-dimensional tables**

 2. As with the F and the t Test Statistics, **there is a different Chi-Square Distribution for each value of Degrees of Freedom** (df).
   - In each case, **the Distribution's Mean is equal to the Degrees of Freedom ($\mu = $ df).**
   - **For larger values of Degrees of Freedom:**
     ○ **the Distributions move to the right**

> o **they become more symmetrical**
> o **Critical Values increase** (move to the right)
> o **the Variances increase** (the Spread becomes wider).

The formula for Degrees of Freedom (symbol df or the Greek letter v) **varies with the Chi-Square test being used.**

Chi-Square Test	df	Explanation
for Goodness of Fit	$n - 1$	n: # bins, columns
for Independence	$(r - 1)(c - 1)$	# of rows and columns
for Variance	$n - 1$	n: Sample Size

In its simplest form, df is one less than the Sample Size. For all three tests, as the Sample Size increases, df increases.

The formulas for the Probability Density Function (which defines the shapes of the Distribution curves) and the Cumulative Density Function (which measures areas under the curves) are complicated and are rarely used. Tables, spreadsheets, or software are used instead to calculate these Probabilities.

Chi-Square Distributions

As shown in the graphs of three χ^2 Distributions below, for larger values of df (and, thus, larger values of the Mean), the Distributions are stretched to the right, and they become more symmetrical. The Critical Values (which mark the left boundary of the shaded area representing $\alpha = 5\%$ in these 1-sided graphs below) also grow larger as df increases.

df = μ = 3	df = μ = 10	df = μ = 30
χ^2-critical = 7.8	χ^2-critical = 18.3	χ^2-critical = 43.8

As an FYI: The shapes of the χ^2 Distributions are similar to those of F-Distributions, as shown below.

F-Distributions

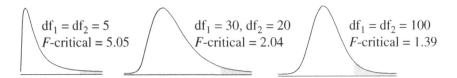

df$_1$ = df$_2$ = 5	df$_1$ = 30, df$_2$ = 20	df$_1$ = df$_2$ = 100
F-critical = 5.05	F-critical = 2.04	F-critical = 1.39

Both the χ^2 and the F Test Statistics are used in tests of the Variance.

 | **3. Furthermore, for All Chi-Square Distributions:**
- the **Mode = df – 2** (for df $\geq$ 3)
- the **Variance = 2df**; **Range: $\chi^2 = 0$ to Infinity**
- they **approach, but never touch the horizontal axis** as they extend to the right.
- they **are not symmetrical** – they are skewed toward the right tail

 | **4.** Since Chi-Square Distributions are not symmetrical, **there are two different Critical Values for a 2-sided test.**

The graphs above showed 1-sided, right-tailed tests. The Cumulative Probabilities (shaded areas) for p or Alpha were calculated only under the right tail of the curves. For 1-sided (either left-tailed <u>or</u> right-tailed) tests, there is only one Critical Value.

For 2-sided tests using the Test Statistics z and t, which have symmetrical Distributions, there is only one Critical Value. That Critical Value is added or subtracted from the Mean.

Since Chi-Square's Distributions are not symmetric, **the areas under the curve at the left tail and the right tail side have different shapes**, for a given value of that area. **So, there are two different Critical Values – an Upper and a Lower – for a 2-sided Chi-Square test.**

Unlike z and t, we do not add or subtract these from the Mean. **The two Critical Values of Chi-Square** produced by tables, spreadsheets, or software **are the final values to be used**.

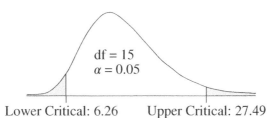

df = 15
$\alpha = 0.05$

Lower Critical: 6.26   Upper Critical: 27.49

If you're looking these up in a 2-sided table, you may need to look up the Critical Values for $\alpha/2$ and for $1 - \alpha/2$. Sometimes two different tables are provided for Upper and Lower Critical Values. Or spreadsheets or software will do this for you.

> **5. Chi-Square is used to analyze Variances via three different Chi-Square Tests: for the Variance, for Independence, and for Goodness of Fit.**

The Test Statistics z and t are used in analyzing Means. Chi-Square and F are used in analyzing Variances.

The three Chi-Square tests use different methods for the different types of analyses involving Variances. Each of these three tests is described individually in one of the three articles which follow this article. Here is a summary:

- ## Chi-Square Test for the Variance

This test **compares the Sample Variance to a value of Variance which we specify.** The value we specify for the Variance could be a target, or a historical value, or anything else. **The test tells us whether there is a Statistically Significant difference between the Sample Variance and the specified Variance.** It is analogous to the 1-Sample t-test for Means.
With this test, the formula for χ^2 is

$$\chi^2 = (n - 1)\frac{\sigma^2}{s^2}$$

where σ^2 is the specified Variance and s^2 is the Sample Variance.

If you are familiar with the t-tests, this Chi-Square Test for the Variance is analogous to the 1-Sample t-test for the Mean, and the F-test is analogous to the 2-Sample t-test:

We Want to Compare	the Statistic	Test to Use
the Calculated value of a Sample Statistic with a value we specify	Variance	Chi-Square Test for the Variance
	Mean	1-Sample t-test
Calculated values of Statistics from two different Samples	Variance	F-test
	Mean	2-Sample t-test

- ## Chi-Square Test for Goodness of Fit

This test compares Observed Counts to Expected Frequencies. For example the table below contains our estimate (Expected Frequencies) for Counts of customers to a bar we are about to open. This is compared with actual Observed Counts.

	Monday	Tuesday	Wednesday	Thursday	Friday	Saturday
Expected Frequencies	102.5	102.5	102.5	102.5	246	164
Observed Counts	98	112	91	102	244	160

The test will tell us whether there is a Statistically Significant difference between our plan and the actual, or whether there is a good fit.

With this test, the formula for χ^2 is the sum, for each cell, of $(O - E)^2$ divided by E.

$$\chi^2 = \sum \frac{(O - E)^2}{E}$$

- **Chi-Square Test for Independence**

This test also uses a 2-dimensional table of data values. And **its formula for χ^2 is the same as for the Goodness of Fit test:**

$$\chi^2 = \sum \frac{(O - E)^2}{E}$$

This test will tell us whether there is an Association or a Statistically Significant difference between two Categorical (aka Nominal) **Variables** (e.g., Gender and Fruit Juice Preference).

Related Articles in This Book: *Test Statistic*; *Distributions – Parts 1– 3*; *Degrees of Freedom*; *Critical Values*; *F*; *Variance*; *Chi-Square Test for Goodness of Fit*; *Chi-Square Test for Independence*; *Chi-Square Test for the Variance*

CHI-SQUARE TEST FOR GOODNESS OF FIT

This article builds on the content of the article, "Chi-Square: Test Statistic and Distributions".

Summary of Keys to Understanding

 1. The Chi-Square (χ^2) Test for Goodness of Fit is a 1-way test of a Categorical (aka **Nominal**) **Variable.**

 2. The Test can be used to determine whether (Observed) Sample data:
- **Fit a specified set of values** (e.g., our estimate)
- **Fit a specified Discrete or Continuous Distribution** (e.g., are the data Normal?)

 3. The Test determines whether there is a Good Fit between Observed (O) Counts from data and Expected (E) Frequencies which we have specified or a Distribution.

	Monday	Tuesday	Wednesday	Thursday	Friday	Saturday
Expected Frequencies	102.5	102.5	102.5	102.5	246	164
Observed Counts	98	112	91	102	244	160

 4. Null Hypothesis (H_0): There is no Statistically Significant **difference between the Observed Counts and the Expected Frequencies.** Therefore, **there is a Good Fit.**

 5. Test Statistic: $\chi^2 = \sum \dfrac{(O - E)^2}{E}$

Critical Value: determined by α, and the Degrees of Freedom, df. df $= n - 1$, where n is the number of categories of the Variable.

Explanation

> 1. The Chi-Square (χ^2) Test for Goodness of Fit is a 1-way test of a Categorical (aka Nominal) Variable.

"1-way" means that a single Variable is involved. So this test is less complicated than the Chi-Square test for Independence, which involves two Variables.

The values of a numerical Variable like height or weight or temperature are numbers.

The values of a Categorical Variable are names of categories.

Variables			
Numerical examples		Non-Numerical/ Categorical examples	
Variable	Example Values	**Variable**	**category names**
weight	102.4 kilograms	gender	female, male
temperature	98.6 degrees F	process	before, after

The numerical data for a Categorical Variable are the Counts within each category.

In the example we'll be using for this article, there are
1 Categorical Variable: Day of the Week
6 Categories: Monday, Tuesday, Wednesday, Thursday, Friday, Saturday
1 Count for each category: 106, 112, 91, 102, 211, 143

Categories	Monday	Tuesday	Wednesday	Thursday	Friday	Saturday
Counts for each category	106	112	91	102	211	143

> 2. The Test can be used to determine whether (Observed) Sample data:
> – **Fit a specified set of values** (e.g., our estimate)
> – **Fit a specified Discrete or Continuous Distribution** (e.g., are the data Normal?)

A number of statistical tests assume the data are Normal. So, before using those tests, one must make sure. The Chi-Square Goodness of Fit test can be used as an alternative to the Anderson-Darling test for Normally distributed data. It is unusually versatile in that it can also be used on other

Continuous Distributions as well as on Discrete Distributions such as the Binomial or Poisson.

This test is also useful in statistical modelling to determine whether a specified Model fits the data.

 3. **The Test determines whether there is a Good Fit between Expected (E) Frequencies which we have specified or a Distribution and Observed (O) Counts from data.**

Expected: We estimate or state or hypothesize numbers that we have reason to expect would be borne out by any Sample data. It's simpler if we can state our expected numbers in actual Counts. But it is often the case that we need to deal in Proportions (in decimal format) or percentages.

For example, let's say we're about to open a new bar, and we want to plan staffing levels. We know from past experience that the number of customers varies by day of the week. We don't know how many customers to expect in a week, but we can estimate what percentages to expect each day. We will be closed on Sundays. Here's what we expect:

Expected Percentages

Monday	Tuesday	Wednesday	Thursday	Friday	Saturday	Total
12.5%	12.5%	12.5%	12.5%	30%	20%	100%

Anticipating that we will be doing a statistical test on the validity of the model represented by these percentages, we select Alpha = 5%, which gives us a 95% Level of Confidence in the test. We opened the bar, and we counted customers for 6 days. We observed the following:

Observed Counts

Monday	Tuesday	Wednesday	Thursday	Friday	Saturday
106	112	91	102	211	143

"Observed" Counts are actual numbers from the Sample data. Counts, by definition, are always non-negative integers (i.e., 0, 1, 2, 3 …).

Now, to compare Observed Counts to Expected percentages, **do not convert Observed Counts from the Sample data into Proportions or percentages**. This is because Counts contain information related to Sample Size, and that information would get lost in converting to Proportions or percentages. **Instead, if needed, multiply the Expected Percentages by the Total Count to get Expected Frequencies.**

	Monday	Tuesday	Wednesday	Thursday	Friday	Saturday
Expected Frequencies	102.5	102.5	102.5	102.5	246	164
Observed Counts	106	112	91	102	211	143

These **Expected Frequencies don't have to be integers,** like Counts do.

The **Chi-Square Test for Goodness of Fit has certain minimum size requirements (test Assumptions).**

- **Every Expected Frequency must be 1 or greater**
- **and no more than 20% of the Expected Frequencies can be below 5.**

If either of these Assumptions are not met, increasing the Sample Size will often help.

Always plot the data. If the plot shows that there is obviously no fit, then do not proceed with the test.

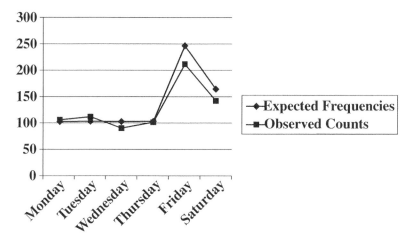

This plot looks like there is a fit. But to determine whether that fit is "good enough" to meet our desired Level of Confidence, we'll need to perform the Chi-Square Test for Goodness of Fit. If we're going to use Hypothesis Testing, we'll need to state a Null Hypothesis for this test.

 4. Null Hypothesis (H_0): There is no Statistically Significant **difference between the Observed Counts and the Expected Frequencies.** Therefore, **there is a Good Fit.**

This may be confusing at first, because we are used to having H_0 be a statement of nothingness – there is no difference, no change, or no effect. And now we are saying that it does mean something – there is a Good Fit.

But, the thing to remember is that **a <u>Good Fit</u> means the same as <u>no</u> <u>difference.</u>**

No difference ...

... means a Good Fit.

The following table will help reinforce this.

If the Test Results are:	$p \leq \alpha$ and $\chi^2 \geq$ χ^2-critical	$p > \alpha$ and $\chi^2 <$ χ^2-critical
	then	then
Is there a difference?	Yes	No
Is there a Good Fit?	No	Yes
Null Hypothesis	Reject	Fail to Reject (Accept)

> **5. Test Statistic:** $\chi^2 = \sum \dfrac{(O - E)^2}{E}$
>
> **Critical Value: determined by α, and the Degrees of Freedom, df.**
>
> **df $= n - 1$,** where n is the number of categories.

<u>**For each category**</u> (each day of the week in our example), **the Test...**

- Subtracts the Expected Frequency from the Observed Count: $O - E$
- Squares it to make it positive: $(O - E)^2$
- Divides by the Expected Value $(O - E)^2/E$
- Sums these for all the cells to get the Chi-Square **Test Statistic,** χ^2

$$\chi^2 = \sum \frac{(O - E)^2}{E}$$

The numerator $(O - E)^2$ in the Test Statistic formula makes it clear why **the larger the difference between O and E, the larger the value of the Chi-Square Test Statistic.** And, therefore

- the farther to the right χ^2 is on the graph below, thus
- the more likely it is that $\chi^2 > \chi^2$-critical, thus
- the more likely it is that χ^2 is in the Not a Good Fit range
- the more likely it is that there is <u>not</u> a Good Fit

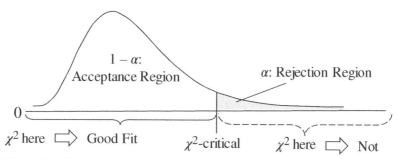

Next, the Test …

- Calculates the Degrees of Freedom: **df $= k - 1$,** where k is the number of categories of the Variable (the 6 days of the week, in our example)
- Uses df to identify the appropriate Chi-Square Distribution
- Uses α and the Distribution to calculate the Critical Value, χ^2**-critical**
- Compares χ^2 to χ^2-critical. Or equivalently, compares p to α

If $\chi^2 \geq \chi^2$-critical, or equivalently, $p \leq \alpha$,

- **There <u>is</u> a Statistically Significant difference.**
- **There <u>is not</u> a Good Fit.**
- **Reject H$_0$, the Null Hypothesis.**

If $\chi^2 < \chi^2$-critical, or equivalently, $p > \alpha$,

- **There <u>is not</u> a Statistically Significant difference.**
- **There <u>is</u> a Good Fit.**
- **Fail to Reject** (Accept) **H$_0$, the Null Hypothesis.**

In our bar day-of-the-week example,
The formula for Chi-Square gives us a Test Statistic $\chi^2 = 9.96$
Since there are 6 columns (Monday – Saturday) df $= 6 - 1 = 5$
This gives us the Critical Value, χ^2-critical $= 11.07$
The Test Statistic value (χ^2) of 9.96 is less than the Critical Value (χ^2-critical) of 11.07. (This result is equivalent to $p > \alpha$). In the diagram above, χ^2 is in the white Acceptance (of the Null Hypothesis) Region. We

Accept (Fail to Reject) the Null Hypothesis which states there is no Statistically Significant difference. We noted earlier, and this diagram reminds us, that **that means there is a Good Fit**.

We might wonder – especially if there were a lot of rows and columns in the table – **which particular values made the difference between Good Fit and not. The test results don't tell us that**. However, We could look at the contribution of each cell to the value of χ^2. That would be:

$$\frac{(O - E)^2}{E}$$

If needed, a Pareto chart could be used to help determine which cells contributed enough to tip the scales. (See the article *Charts, Graphs, Plots – Which to Use When*.)

Related Articles in This Book: *Chi-Square: the Test Statistic and Its Distributions*; *Test Statistic*; *Alpha, p-Value, Test Statistic, and Critical Value – How They Work Together*; *Degrees of Freedom*; *Hypothesis Testing*; *Null Hypothesis*; *p-Value, p*; *Reject the Null Hypothesis*; *Fail to Reject the Null Hypothesis*; *Chi-Square Test of Independence*; *Chi-Square Test of the Variance*; *Charts, Graphs, Plots – Which to Use When*

CHI-SQUARE TEST FOR INDEPENDENCE

This article builds on the content of the article, "Chi-Square: Test Statistic and Distributions".

Summary of Keys to Understanding

 1. The Test Statistic, Chi-Square (χ^2), can be used to test whether two Categorical (aka "Nominal") **Variables are Independent or Associated.**

 2. If two **Categorical Variables** are **Independent**, then the **Observed Frequencies** (Counts) of the different values of the **Variables** should be **Proportional**.

Juice Study: Proportions are the same, so the Variables, Gender and Juice are Independent					Ice Cream Study: Proportions are very different, so the Variables, Gender and Ice Cream are Associated (not Independent).				
	female		male			female		male	
	Count	Proportion	Count	Proportion		Count	Proportion	Count	Proportion
apple	28	0.35	14	0.35	chocolate	48	0.48	16	0.20
grape	12	0.15	6	0.15	strawberry	28	0.28	40	0.50
orange	40	0.50	20	0.50	vanilla	24	0.24	24	0.30
Total	80	1.00	40	1.00	Total	100	1.00	80	1.00

 3. Using a Contingency Table, calculate the difference between the Observed Count and the Expected Frequency for each cell. The Test Statistic Chi-Square then distills all this information into a single number:

$$\chi^2 = \sum \frac{(O - E)^2}{E}$$

 4. The Test then uses the appropriate Chi-Square Distribution to calculate χ^2-critical and p. Null Hypothesis: there is no Association, that is, the Variables are Independent.

Explanation

 1. **The Test Statistic, Chi-Square (χ^2), can be used to test whether two Categorical** (aka "Nominal" or "Attributes") **Variables are Independent or Associated.**

To make things a little clearer, in this article, the Variables are shown in upper case, while their values (the categories) are shown in lower case.

What is a Categorical Variable?

A Categorical Variable is one whose values are names of categories.

The Table below has two Categorical Variables, Gender and Ice Cream.

The Categorical Variable, Gender, has two values: the category names "female" and "male."

The Categorical Variable, Ice Cream flavor, has three values: the category names "chocolate," "vanilla," and "strawberry."

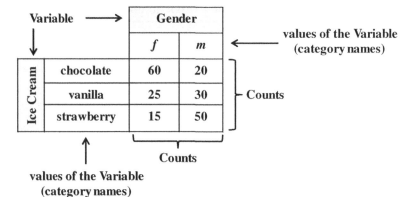

Categorical Variables are classified as "non-numerical" Variables, because their values are names, not numbers. However, we get numbers by counting items within categories.

In this table, Count data are recorded for each intersection of categories from the two Variables. For example, at the intersection of the female column and the chocolate row, there were 60 people who were both female and preferred chocolate.

Independent vs. Associated

Two or more Variables are **Independent** of one another if there is **no relationship, Association, or Correlation between their values**.

Examples:

Independent Variables: gender and day of birth – there is not a higher Proportion of boys born on Tuesdays vs. Thursdays, for example

Associated Variables: gender and types of clothes worn – many more females wear skirts than males.

Independence is determined between Variables not between values (category names) of the Variables. For example, "Is Ice Cream flavor preference dependent on Gender?" – Not: "Is "female" dependent on "male" or is "male" dependent on "strawberry.""

For Numerical Variables, "Correlation," is the opposite of Independence, and there are a number of types of Correlation analyses available. **For Categorical Variables, the term "Association" is used instead of "Correlation,"** and the Chi-Square test of Independence is used.

> **2. If two Categorical Variables are Independent, then the Observed Frequencies (Counts) of the different values of the Variables should be Proportional.**

(In the first few tables shown here, "Frequency" means the same as Count. Later, we'll be doing some calculations, which result in non-integer "Frequencies.") In this example, there are two Categorical Variables: Gender and Juice preference. The Variable, Gender, has two possible values – the category names "male" and "female." The Categorical Variable, Juice, has three possible values – the category names "apple," "grape," and "orange."

Q: Are the Variables, Gender and Juice, Independent? Put another way: Does your Gender have no effect on your choice of fruit Juice?

Let's say we surveyed 80 females and recorded their Juice preferences in the table below.

This is the beginning of a **Contingency Table**. It is so named because we're trying to determine whether the Counts for one of the two Categorical Variables (e.g., Juice) are *contingent* on (dependent on) the value of the other Variable (e.g., Gender).

If Gender has no effect on fruit Juice preference (that is, if these two Categorical Variables are **Independent**), then we would expect that the preferences for males would follow a similar 35%/15%/50% pattern.

		Gender	
		female	
		Count	Proportion
Juice	apple	28	0.35
	grape	12	0.15
	orange	40	0.50
	Total	80	1.00

So, if we surveyed 40 males, and if the Variables, Gender and Juice <u>were</u> Independent, one would expect the Proportions for males and females to be close to equal. Half as many males are in the Sample, so we would expect half as many males for each value of the Juice Variable. So, **the Proportions or percentages would be the same if the two Variables,** Gender and Juice preference, **are Independent** – that is, if Gender has no effect on Juice preference.

Juice Study: Proportions are the same, so the Variables Gender and Juice are Independent					
		Gender			
		female		male	
		Count	Proportion	Count	Proportion
Juice	apple	28	0.35	14	0.35
	grape	12	0.15	6	0.15
	orange	40	0.50	20	0.50
	Total	80	1.00	40	1.00

In terms of Counts, there are half as many males as females in total, and half as many for each type of juice. The corresponding Proportions are identical for male and female. So, from the table above, we conclude that there is <u>no Association</u> between gender and fruit juice preference: the Variables gender and juice are <u>Independent</u>.

Let's see if the same can be said for ice cream. We conduct a new study with 100 females and 80 males and get the following result:

Ice Cream Study: Proportions are very different, so the Variables Gender and Ice Cream are Associated (not Independent).					
		Gender			
		female		male	
		Count	Proportion	Count	Proportion
Ice Cream	chocolate	48	0.48	16	0.20
	strawberry	28	0.28	40	0.50
	vanilla	24	0.24	24	0.30
	Total	100	1.00	80	1.00

Clearly, females preferred chocolate by a wide margin, and males strongly preferred strawberry. So, gender and ice cream preference are <u>Associated, not Independent</u>.

Most of the time, however, things are not so clear-cut. What if the Proportions in the Gender/Juice Contingency Table were not the same, but were close? How close is close enough? **We need to use the Chi-Square test of Independence to determine whether the differences in the Proportions are large enough to be Statistically Significant.** If so, the Variables are Associated, not Independent.

> 3. **Using a Contingency Table, calculate the difference between the Observed Count and the Expected Frequency for each cell. The Test Statistic Chi-Square then distills all this information into a single number:**
>
> $$\chi^2 = \sum \frac{(O - E)^2}{E}$$

Important: the Expected Frequencies in each cell must be 5 or larger. Otherwise, our Sample Size is not big enough to conduct the test, or our categories are defined too narrowly.

In our earlier Gender–Juice example, the Proportions were identical for female and male. That, of course, is not usually the case. Let's say a cafeteria offers a choice of fish, meat, or vegetarian meals. Is the choice of Meal influenced by Gender? We collect the following Observed Counts on the selections of 90 females and 100 males. Thus, the Sample Size $n = 190$.

Observed Counts

		Gender		
		female	male	Row Total
Meal	fish	26	32	58
	meat	29	44	73
	vegetarian	35	24	59
	Column Total	90	100	190

We can see one obvious difference: the first choice was vegetarian for females and meat for males. But, both Genders chose meat over fish. And, the picture is complicated by the fact that there are different total numbers for the two Genders.

So, just looking at the data is insufficient to tell us whether we have a Statistically Significant difference between male and female choices. However, from this data, we <u>can</u> calculate the value of the Chi-Square Test Statistic which we can use in the Test for Independence.

The question is <u>how likely</u> is it that we would get these numbers if there were no difference Associated with Gender (that is, if the two Variables were Independent)? It's a question of Probabilities, so, we can make use of the ...

Multiplicative Law of Probability: the Probability of both of two things happening is the product of their individual Probabilities.

In our table of data above, each cell is the Count of two things happening. For example, the upper left Count of 26 represents female and fish. Each of these two things have their own Probabilities.

Prob(female) = total Count of females divided by total Count of people polled = 90/190.

Note: this is the Probability of the female column.

Prob(fish) = total Count of fish choices divided by Count of all choices = 58/190.

Note: this is the Probability of the fish row.

Therefore, Prob(fish and female) = (90/190) × (58/190) = 0.305 × 0.474 = 0.145.

Note: this is the product of the row and column probabilities for the cell fish–female. So, for a table, the Multiplicative Law of Probability can be restated as:

Expected Probability: The Expected Probability of each cell in a table is the product of the Probability of its row times the Probability of its column.

	f	m	Total	Row Probability
fish	26	32	58	58/190 = 0.305
meat	29	44	73	73/190 = 0.384
vegetarian	35	24	59	59/190 = 0.311
Total Column Probability	90	100	190	1.000
	90/190 = 0.474	100/190 = 0.526		

For each cell, **Expected Probability = Row Probability × Column Probability**

	Expected Probabilities	
	f	m
fish	0.145	0.161
meat	0.182	0.202
vegetarian	0.147	0.163

Expected Frequencies can be calculated from Expected Probabilities:
Expected Frequency = Sample Size × Expected Probability

In, our example, the Sample Size, $n = 190$.

The Expected Frequency is what we would expect to see as data if the Variables were Independent. Note that Frequencies do not need to be integers as Counts do.

	Expected Frequencies	
	f	m
fish	$0.145 \times 190 = \mathbf{27.6}$	$0.161 \times 190 = \mathbf{30.6}$
meat	$0.182 \times 190 = \mathbf{34.6}$	$0.202 \times 190 = \mathbf{38.4}$
vegetarian	$0.147 \times 190 = \mathbf{27.9}$	$0.163 \times 190 = \mathbf{31.0}$

=> **Important:** Although, within the calculations there are comparisons of Proportions, **do not convert the Count data to Proportions or percentages** before entering the data into the table and doing the calculations. **Count data contains information on the Sample Size**, and that information is lost when you convert to Proportions. Likewise, **do not round the Count data.**

In the formula for calculating the Chi-Square Test Statistic from the data, the O represents – for each cell – the Observed Count from the data. The E represents the calculated Expected Frequency for that cell.

For each cell,

- Subtract the Expected Value from the Observed Value: $O - E$
- Square it to make it positive: $(O - E)^2$
- Divide by the Expected Value $(O - E)^2/E$

Then, sum these for all the cells to get the **Chi-Square Test Statistic**, which distills all the values in the Sample into one number. This number **summarizes how much the Sample varies from one which would have Independent Variables.**

$$\chi^2 = \sum \frac{(O - E)^2}{E}$$

This formula makes it clear why **the larger the difference between O and E,**

- **the greater the value of the Chi-Square Test Statistic,** and thus
- **the more likely it is to be greater than the Critical Value,** thus
- **the more likely it is that the Variables are Associated (<u>not</u> Independent).**

It also explains why **this test is always Right-tailed** – only large positive values of χ^2 are in the extremes beyond the Critical Value.

Following the formula, we get a value for χ^2 which summarizes the data and how it differs from what we would expect if the Variables, gender and meal preference, were Independent. In our example, $\chi^2 = 5.24$.

> **4. The Test then uses the appropriate Chi-Square Distribution to calculate χ^2-critical and p. Null Hypothesis: there is no Association, that is, the Variables are Independent.**

If we were doing a Hypothesis Test, we would have stated a Null Hypothesis along with selecting a value for Alpha before collecting the Sample data. A Null Hypothesis, H_0, usually states that there is no difference or no change, or no effect. In this test, we can view it as saying that there the values of one Variable have **no effect** on the Counts in the other Variable.

Null Hypothesis (symbol H_0): **There is no Association; the Variables are Independent.**

The Chi-Square Test

- Uses Alpha(α) and the Distribution to calculate the Critical Value, χ^2-**critical**.
- Uses χ^2 and the Distribution to calculate the **p-value**, p.
- Compares χ^2 to χ^2-critical. Or equivalently, compares p to α.

    If $\chi^2 \geq \chi^2$-**critical**, or equivalently, $p \leq \alpha$,
    The variables are **not Independent**. H_0, Reject the Null Hypothesis.

    If $\chi^2 < \chi^2$-**critical**, or equivalently, $p > \alpha$,
    The variables **are Independent**. Fail to Reject (Accept) the Null Hypothesis, H_0.

There is a different Chi-Square Distribution for each value of Degrees of Freedom. For a Contingency Table with r rows and c columns, calculate the Degrees of Freedom as:

df $= (r - 1) \times (c - 1)$ where r is the number of rows and c is the number of columns.
In our meal preference example, df $= (3 - 1) \times (2 - 1) = 2$

As we explained earlier, the test is right-tailed. For a right-tailed test with df $= 2$ and Alpha selected to be 0.05, the Critical Value χ^2-critical is 5.99. (This can be found in reference tables or via software.)

For our example, the following table is how one software package summarized the results:

	f	m	Total		f Expected	m Expected	Chi Sq-Crit	5.99
fish	26	32	58		27.47	30.53	Chi-Sq	5.24
meat	29	44	73		34.58	38.42	p	0.07
vegetarian	35	24	59		27.95	31.05	α	0.05
Total	90	100			**Conclusion: Variables are Independent**			

(You may have noticed that the f Expected and m Expected numbers calculated by software in the table above differ slightly from the Expected Frequencies table we calculated earlier. This is because, in the previous tables, we rounded several times in order to display interim numbers without overly long strings of decimals.)

$\chi^2 < \chi^2$–**critical and, equivalently $p > \alpha$, so we Fail to Reject (that is, we Accept) the Null Hypothesis of no Association.** So, the Variables, Gender and Meal preference, <u>are</u> Independent.

The graph below is a closeup of the right tail of the χ^2 Distribution. The shaded area representing Alpha is sometimes called the Rejection Region, and the white area is called the Acceptance Region. We can see that χ^2 is in the Acceptance Region.

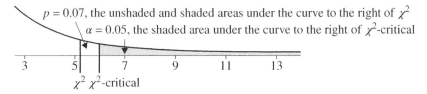

$p = 0.07$, the unshaded and shaded areas under the curve to the right of χ^2
$\alpha = 0.05$, the shaded area under the curve to the right of χ^2-critical

χ^2 χ^2-critical

We might wonder – especially if there were a lot of rows and columns in the table – which particular cells made the difference between Independence and Association. Some software packages provide this as part of the test output. Or, we could calculate the cell values individually using the formula:

$$\frac{(O - E)^2}{E}$$

A Pareto Chart could then help us see which cells contributed enough to tip the scales. See the article *Charts, Graphs, Plots – Which to Use When.*

Related Articles in This Book: *Variables*; *Chi-Square: Test Statistic and Distributions*; *Degrees of Freedom*; *Alpha, p-Value, Test Statistic, and Critical Value – How They Work Together*; *Hypothesis Testing*; *Null Hypothesis*; *Reject the Null Hypothesis*; *Fail to Reject the Null Hypothesis*; *Chi-Square Test for Goodness of Fit*; *Chi-Square Test for the Variance*

CHI-SQUARE TEST FOR THE VARIANCE

Builds on the content of the article, "Chi-Square: the Test Statistic and its Distributions".

> The Chi-Square Test for the Variance compares the Variance calculated from the Sample data to a value for the Variance which we specify.

The Test can be

- **1-sided** (1-tailed left or right) **or 2-sided** (2-tailed).
- framed as a **Confidence Interval** test **or** a **Hypothesis** test.

The specified Variance can be, for example,
a **target value** of a quality measurement,
the **historical value** of a Process Variance,
an **estimate** of a Population or Process Variance.

The Chi-Square test for the Variance is non-Parametric; it can be used with non-Normal variables and data.

Note: the F-test is also a test involving Variances. But the F-test compares Variances from two Samples representing two different Populations or Processes. Also the F-test is Parametric – it assumes the data Distribution is approximately Normal.

If you're familiar with the t-tests, the Chi-Square Test for the Variance is analogous to the 1-Sample t-test, while the F-test is analogous to the 2-Sample t-test.

Procedure for Chi-Square Test for the Variance

0. If Hypothesis Testing is being used, we would state the **Null Hypothesis** (H_0).

- 2-sided: H_0: Population or Process Variance = specified Variance.
- 1-sided, left-tailed: H_0: Population or Process Variance $\geq$ specified Variance.
- 1-sided right-tailed: H_0: Population or Process Variance $\leq$ specified Variance.

where σ^2 is the unknown Population or Process Variance

1. **Select a** value for the **Significance Level**, usually $\alpha = 0.05$.
2. **Collect a Sample of data**, size n.

3. **Determine the Degrees of Freedom; df** $= n - 1$.

4. Use a spreadsheet, software, or table to **calculate the Critical Value(s). Inputs are** α **and df.** A 2-sided test would have two Critical Values. As noted in the article, *Chi-Square – the Test Statistic and its Distributions*, these two would not be equal, since the Chi-Square Distribution is not symmetrical.

5. **From the Sample data, calculate** the value of **the Chi-Square Test Statistic,** χ^2, and/or the Probability, p. Either one contains the same information, since p is the Cumulative Probability of all values beyond the Test Statistic.

$$\chi^2 = (n - 1)\,\frac{\sigma^2}{s^2}$$

where s is the Standard Deviation calculated from the Sample and σ^2 is the specified Variance.

6. **Compare** either χ^2 **to** χ^2**-critical(s) or** compare p **to** α. The comparisons are statistically identical, since α (for 1-sided tests or $\alpha/2$ for 2-sided tests) is the Cumulative Probability for a Critical Value.

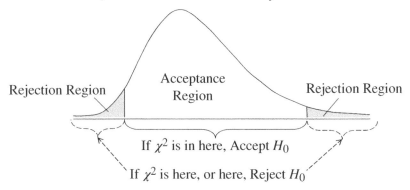

The farther out along either tail, the smaller the Probabilities get. **If the Test Statistic** χ^2 (which was calculated from the Sample data) **is more extreme** (farther out along a right or left tail) **than a Critical Value** (equivalently, if $p \leq \alpha$), we conclude that the Probability is very small that this result is due only to chance. Thus, **there is a Statistically Significant difference (Reject the Null Hypothesis, H_0)** between the specified Variance and the Variance of the Population or Process from which our Sample was taken.

If the Test Statistic χ^2 **is not** more extreme than a Critical Value (equivalently, if $p > \alpha$), then **there is not a Statistically Significant difference (Fail to Reject H_0/Accept H_0).**

The **Confidence Interval is between the Critical Values.** Confidence Intervals for Variances and Standard Deviations are wider than those for Means, because estimates of Means are more accurate than estimates of Variances or Standard Deviations.

Related Articles in This Book: *Variance*; *Chi-Square – the Test Statistic and Its Distributions*; *Degrees of Freedom*; *Alpha, p-value, Test Statistic and Critical Value – How They Work Together*; *Hypothesis Testing – Parts 1 and 2*; *Null Hypothesis*; *Reject the Null Hypothesis*; *Fail to Reject the Null Hypothesis*; *Chi-Square Test for Goodness of Fit*; *Chi-Square Test for Independence*; *p, t, and F: ">" or "<" ?*

CONFIDENCE INTERVALS – PART 1 (OF 2): GENERAL CONCEPTS

Summary of Keys to Understanding

 1. **There are two main methods in Inferential Statistics:**
   - **Hypothesis Testing, which produces a point estimate**
   - **Confidence Intervals, which produces an Interval Estimate**

 2. **The selected Level of Significance, α, is plotted on the Distribution of a Test Statistic to determine the Critical Value.**
   We then convert a Critical Value into units of the data Variable (x) in order to get the Limits which define the Confidence Interval.

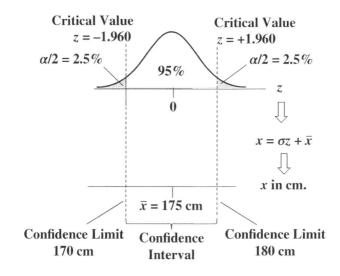

Critical Value
$z = -1.960$
$\alpha/2 = 2.5\%$

Critical Value
$z = +1.960$
$\alpha/2 = 2.5\%$

95%

z

0

$x = \sigma z + \bar{x}$

x in cm.

$\bar{x} = 175$ cm

Confidence Limit
170 cm

Confidence
Interval

Confidence Limit
180 cm

 3. **The Margin of Error (MOE) is one-half the width of 2-sided Confidence Interval. Three factors affect the size of the MOE: Sample Size (n), Level of Significance (α), and Standard Deviation (σ or s).**

 4. **A Confidence Bound defines the boundary of a 1-sided Confidence Interval.**

Explanation

In Inferential Statistics, we calculate a Sample Statistic, e.g., its Mean or Standard Deviation, **as an estimate of the corresponding property of the Population or Process** from which the Sample was collected (e.g., the Population or Process Mean).

> **1. There are two main methods in Inferential Statistics:**
> - **Hypothesis Testing, which produces a point estimate**
> - **Confidence Intervals, which produces an interval estimate**

In Hypothesis Testing, the estimate is a point value: "With a 95% Level of Confidence, we can say that the Population Mean is 175 cm."

In the Confidence Interval method, the estimate is an interval: **"With a 95% Level of Confidence, we can say that the Population Mean is between 170 and 180 cm."**

So what's the difference? Statistically, none. Both go through similar calculations, they just present the conclusions in different ways.

- **Hypothesis Testing may be more useful when you want a Yes or No answer about the Statistical Significance of an observed difference, change, or effect.** For example,
  - Is there a Statistically Significant difference between our school's test results and the national average?
  - Has there been a Statistically Significant change in our Process's defect rate?
  - Does the training have a Statistically Significant effect?

Note that these all involve comparisons. In the last case, the comparison is with zero.

- **Confidence Intervals may be more useful when you want an estimate of a Population or Process Parameter.**

For example, What is the average height of adult males in the Population?

Confidence Intervals also provide more information, namely, the Margin of Error. More on that later in this article.

- **Confidence Intervals may be less confusing.**

Hypothesis Testing, as detailed in this book's articles on the subject, can be confusing – even for experienced practitioners. For example, you have to be prepared to deal with confusing language like this triple negative: "We Fail to Reject the Null Hypothesis."

The output of a Confidence Interval analysis (see the Part 2 article) usually includes a graph. Visuals are usually helpful in improving understanding.

> **2. The selected Level of Significance, α, is plotted on the Distribution of a Test Statistic to determine the Critical Value.**
>   **We then convert a Critical Value into units of the data Variable (x) in order to get the Confidence Limits which define the Confidence Interval.**

The first thing to do – before collecting a Sample of data – is to select a Level of Significance, α. Alpha is the level of risk of an Alpha Error (also known as Type I Error or False Positive) which we can tolerate. The Level of Confidence is calculated as 1 – the Significance Level. Most often $\alpha = 5\%$ is chosen, giving a 95% Level of Confidence.

We select a **Test Statistic** to use – such as z, t, F, or Chi-Square (χ^2) – appropriate to the purpose of the analysis and the Sample Size (n). Since z – used for estimating the Mean – is the simplest, we'll use it in this Part 1 article.

A Test Statistic has a Probability Distribution curve with known Probabilities. That is, for any value of the Test Statistic, the Probability of that value occurring is known. And the Cumulative Probabilities of all values in a range of the Test Statistic can be calculated.

Alpha is a Cumulative Probability represented by a shaded area or areas under the curve.

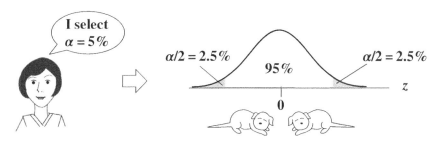

The illustration above is for a 2-sided (2-tailed) analysis, using the Standard Normal Distribution and its associated Test Statistic, z. z is the horizontal axis Variable. The vertical axis is the Probability of a value of z. The 5% shaded area representing Alpha is split between the two tails.

The value of the Test Statistic which forms a boundary for the shaded area(s) representing **Alpha** is a called a **"Critical Value."** Critical Values for common choices of Alpha can be found in tables or calculated with software. The Critical Value of z for $\alpha/2 = 2.5\%$ is 1.960.

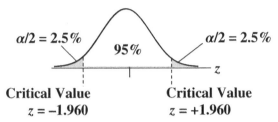

We needed to use a Test Statistic in order to find the values on the horizontal (z) axis which correspond to the Cumulative Probability, Alpha/2. These are the Critical Value + or − the Mean.

Critical Values are in the units of the Test Statistic (z in this example). **A Confidence Interval is in the units of the data Variable, x.**

To make use of Critical Values in the real world, we need to convert the information into the units of the data Variable, x – centimeters in this case.

In doing so, **we convert Critical Values into Confidence Limits.**

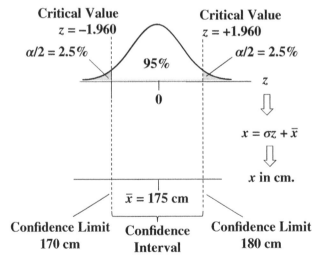

The formula for converting a value of z into a value of x is

$$x = \sigma z + \bar{x}$$

... where σ is the Standard Deviation of the Population or Process and $\bar{x}$ is the Sample Mean. If we don't know the value of σ, then we would have to use the Test Statistic *t* instead of *z*, and we would use the Standard Deviation of the Sample, *s*, instead of σ.

How to interpret the Confidence Interval in this example:

- If we were to calculate Confidence Intervals from many Samples of the same size, 95% would contain the true Mean.
- With a 95% Level of Confidence, we can say that the Population Mean is between 170 and 180 cm, or
- With a 95% Level of Confidence, we can say that the Population Mean is 175 cm plus or minus an MOE of 5 cm.

Most texts caution against saying, "The Confidence Interval of 170 – 180 cm has a 95% <u>Probability</u> of including the Population Mean." This is because the concept of Confidence is wrapped up in a method which involves taking Samples, calculating a Test Statistic, and using its Distribution and Alpha to calculate Critical Values. It is not a direct calculation of Probabilities. However other experts say that the distinction has little practical consequence.

 3. The Margin of Error (MOE) is one-half the width of 2-sided Confidence Interval. Three factors affect the size of the MOE: Sample Size (*n*), Level of Significance (α), and Standard Deviation (σ or *s*).

Let's continue with our example of the Mean and the *z*-Distribution, since it's the simplest case. The formula for MOE is

$$\text{MOE} = \frac{\sigma \, (z\text{-critical})}{n}$$

Alpha, the Level of Significance, determines the value of *z*-critical. *n* is the Sample Size.

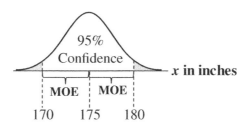

"With a 95% Level of Confidence, we estimate the Mean height to be 175 cm, with an MOE of plus or minus 5 cm."

See the article, *Margin of Error* for more.

> **4. A Confidence Bound defines the boundary of a 1-sided Confidence Interval.**

To this point, we've been discussing 2-sided Confidence Intervals. Alpha is split into two shaded areas, each representing $\alpha/2$. A 2-sided Confidence Interval places both an Upper Confidence Limit and a Lower Confidence Limit on the estimated value of the Parameter (e.g., the Population Mean).

   A 1-sided (1-tailed) **Confidence "Interval,"** however, **places either an Upper Confidence Bound or a Lower Confidence Bound on the estimate.**

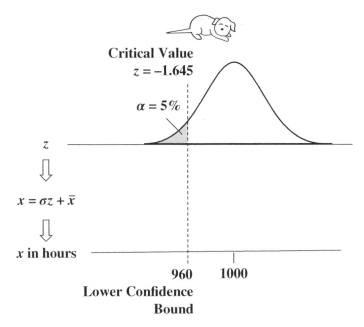

Critical Value
$z = -1.645$

$\alpha = 5\%$

z

$x = \sigma z + \bar{x}$

x **in hours**

960   1000

Lower Confidence Bound

There is some disagreement on whether the term "Confidence Interval" should be used for 1-sided analyses. The dictionary defines "interval" as a "space between two things." In the case of 1-sided Confidence Intervals, one of the two things is missing, because the range of Confidence extends to infinity in the direction opposite the Bound.

This presents a difficulty with using the concept of Margin of Error (MOE). The MOE is defined as half the width of the Confidence Interval, and it would be half of infinity – still infinitely wide – for a 1-sided analysis. However, if we want a 1-sided analysis, we really are not interested in an MOE on both sides of the center of an estimate.

In a 1-sided Confidence Interval, Alpha is not split into two halves, but it is entirely under the left or right tail of the Distribution. And the unshaded Level of Confidence area – under which lies Confidence Interval – extends to infinity in the other direction.

That is important to remember if you are using software which doesn't distinguish between 2-sided and 1-sided Confidence Intervals. You can use the Upper Confidence Limit in a 2-sided interval to give you an Upper Confidence Bound, for example, but you must input an Alpha which is twice that of which you want to see under the single tail of the Distribution. For example, enter $\alpha = 10\%$. Then the software will divide that in two, putting $\alpha/2 = 5\%$ under both the left tail and the right tail. You are ignoring the left tail, so you will have the correct value for Alpha under the right tail. That will give you a Upper Confidence Limit which you can use as your Upper Confidence Bound for a right-tailed analysis.

Related Articles in This Book: *Confidence Intervals – Part 2: Some Specifics*; *Hypothesis Testing – Parts 1 and 2*; *Critical Value*; *Alpha (α)*; *Test Statistic*; *z*; *Margin of Error*; *Alpha and Beta Errors*

CONFIDENCE INTERVALS – PART 2 (OF 2): SOME SPECIFICS

Summary of Keys to Understanding

1. Purpose	Test Statistic	Confidence Interval (CI) Formula
Estimate **Mean** (when σ is known and $n > 30$)	z	$CI = \bar{x} \pm \sigma Z_{\alpha/2}/\sqrt{n}$
Estimate **Mean**	t	$CI = \bar{x} \pm s t_{\alpha/2}/\sqrt{n}$
Estimate **Proportion**	z	$CI = \hat{p} \pm z_{\alpha/2}\sqrt{\hat{p}(1-\hat{p})/n}$
Estimate **Variance**	χ^2	$CI = s^2 \pm (n-1)s^2/\chi^2_{\alpha/2}$

2. **The form of the output of Confidence Intervals analysis varies with the type of analysis** (e.g., Regression, *t*-test) **and the type of software.**

95% Confidence Interval for Mean
   75.551   81.915
95% Confidence Interval for Median
   75.229   81.771

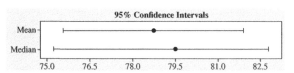

3. **Confidence Intervals which do not overlap indicate a Statistically Significant difference.** Experts disagree on whether the opposite is always true.

Individual 95% CIs for Mean

Pressure	Mean	
High Pressure	34.2667	(-----*-----)
Low Pressure	23.0667	(----*----)
Med Pressure	25.6000	(----*-----)

```
    ---------|---------|---------|---------|
            25        30        35        40
```

4. **For Processes, use Confidence Intervals only if Control Charts show that the Process is under Control.**

Explanation

1. Purpose	Test Statistic	CI Formula
Estimate **Mean** (when σ is known and $n > 30$)	z	$CI = \bar{x} \pm \sigma z_{\alpha/2}/\sqrt{n}$
Estimate **Mean**	t	$CI = \bar{x} \pm s t_{\alpha/2}/\sqrt{n}$

Mean

In the Part 1 article, we used the Test Statistic z to illustrate the concepts, because z is somewhat simpler to use than t. There is only one z Distribution, the Standard Normal Distribution. It is defined by having a Mean (μ) $= 0$ and a Standard Deviation (σ) $= 1$. z can be convenient to use for quick estimates, especially if one has memorized some of its values, such as:
   z for Alpha $= 5\%$ is 1.645, and the corresponding z for $\alpha/2 = 1.96$.
   In the Confidence Interval (CI) formulas in the table above,

   – $\bar{x}$ is the Sample Mean, our Statistic
   – σ, Sigma, is the Standard Deviation of the Population or Process
   – $z_{\alpha/2}$ is the Critical Value of z for $\alpha/2$.
   – n is the Sample Size
   – $t_{\alpha/2}$ is the Critical Value of t for $\alpha/2$.
   – s is the Standard Deviation of the Sample

But z has its limitations. It should only be used only when we know the Standard Deviation (σ, Sigma) of the Population or Process – or if we believe we have a very good estimate of it – **and when the Sample Size (n) is large enough**, say when $n > 30$ (some say when $n > 100$). And we usually don't know Sigma, because if we had exact information about the Population or Process, we wouldn't need to estimate things from Samples.

   Also, the value of z is the same no matter the Sample Size, so it doesn't take into account the improving accuracy of ever-larger Samples. t does. There is a different t Distribution for each different value of $n - 1$, the Degrees of Freedom, df.

   Use t instead of z when Sigma is not known, or when the Sample Size is small. Since accuracy suffers when Sigma is not known, **the Confidence Intervals calculated with t are wider than those calculated with z.** That is, their MOE is larger.

Confidence Intervals for the Mean and Proportion for Non-Normal Data.

In general, the assumption or requirement for using both the t and z Test Statistics is that the data are approximately Normally distributed. **But, for**

$n > 30$ **we can use** t **and** z **for Confidence Intervals for the Mean even if the data are not Normally distributed.** This is due to the Central Limit Theorem, which is explained in the article *Normal Distribution.*

Proportion

Purpose	Test Statistic	CI Formula
Estimate **Proportion**	z	$\mathbf{CI} = \hat{p} \pm z_{\alpha/2}\sqrt{\hat{p}\,(1-\hat{p})/n}$

where,

- $\hat{p}$ is the Sample Proportion
- the Critical Value of z for $\alpha/2$ is denoted by $z_{\alpha/2}$
- n is the Sample Size

Proportions (the decimal equivalent of percentages) are often used with Categorical/Nominal Variables. Such Variables include Counts of things like gender, voters for a particular candidate, favorite fruit juice, etc.

Let's say our Categorical Variable is favorite fruit juice. We survey $n = 200$ people for their favorite and get the following data for the three values (apple, orange, grape) of the Variable.

favorite flavor	Count	Proportion
apple	44	0.22
orange	104	0.52
grape	52	0.26
Total	200	

Confidence Interval for the Proportion of orange as favorite juice:

$$CI = 0.52 \pm 1.96\sqrt{(1-0.52)/200} = 0.52 \pm 0.10$$

Variance

Purpose	Test Statistic	CI Formula
Estimate Variance	χ^2	$\mathbf{CI} = s^2 \pm (n-1)s^2/\chi^2_{\alpha/2}$

Whereas the Sampling Distribution of the Mean and Proportion follow a Normal Distribution, the Sampling Distribution of the Variance follows the Chi-Square (χ^2) Distribution. So the Test Statistic we use is χ^2.

The Central Limit Theorem does not apply to Variances. **Confidence Intervals for the Variance assume that the underlying data follow a Normal Distribution.**

Since the Standard Deviation is defined as the square root of the Variance, **the Confidence Interval for the Standard Deviation is just the square root of the Confidence Interval for the Variance.**

 | **2. The form of the output of Confidence Interval analysis varies with the type of analysis** (e.g., *t*-test, Regression) **and the type of software.**

The following are some examples of Confidence Interval calculated with different statistical software.

Example output from Multiple Linear Regression

	Coefficients	Std Err	*t*-Stat	*p*-Value	Lower 95%	Upper 95%	Lower 99%	Upper 99%
Intercept	−34.750	40.910	−0.849	0.458	−164.944	95.445	−272.702	204.203
House Size	−5.439	21.454	−0.254	0.816	−73.716	62.838	−130.751	119.874
Bedrooms	85.506	15.002	5.700	0.011	37.763	133.249	−2.119	174.132
Bathrooms	77.486	18.526	4.183	0.025	18.529	136.443	−30.721	185.693

"Lower 95%" and "Upper 95%" refer to Lower and Upper Confidence Limits for $\alpha = 5\%$. The Confidence limits are also given for $\alpha = 1\%$.

Example output: CIs for Mean and Median

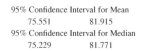

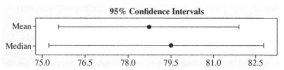

Example output from 1-Sample *t*-test

One-Sample T: Pulse

Variable	N	Mean	StDev	SE Mean	95% CI
Pulse	30	78.73	8.52	1.56	(75.55, 81.92)

Example output from 2-Sample *t*-test

Two Samples are tested to see if they are statistically the same. The difference in their Means is analyzed. **If the zero is within the Confidence Interval,** then there **is no Statistically Significant difference** and (if Hypothesis Testing is used), **the Null Hypothesis** (of no difference) **is Accepted** (that is, we Fail to Reject it).

2-Sample *t* for Furnace 1 vs. Furnace 2 and CI: Furnace 1, Furnace 2

	n	Mean	Std Dev	SE Mean
Furnace 1	40	9.91	3.02	0.48
Furnace 2	50	10.14	2.77	0.39

Difference = mu (Furnace 1) – mu (Furnace 2)
Estimate for difference: – 0.235
95% CI for difference: (−1.450, 0.980)

 | **3. Confidence Intervals which do not overlap indicate a Statistically Significant difference.** Experts disagree on whether the opposite is always true.

If the analysis is comparing Samples from two or more Populations or Processes, it can produce Confidence Intervals for the Mean of each Sample.

If Confidence Intervals do <u>not</u> overlap, then any observed difference in the Sample Statistics is Statistically Significant. Conversely, some experts say that any overlap means no difference. But other experts say there <u>can</u> be a Statistically Significant difference with a slight overlap. In that case, a Hypothesis Test could be used to make sure. In the example below, there is definitely a Statistically Significant difference between High Pressure and the other two treatments. And the overlap for Low and Medium Pressure is sufficiently large to conclude there is not a Statistically Significant difference between them. If there is any doubt, use a 2-Sample *t*-test for two of the Means that are in question.

```
                                    Individual 95% CIs for Mean
Pressure             Mean
High Pressure        34.2667                        (-----*-----)
Low Pressure         23.0667     (----*----)
Med Pressure         25.6000         (----*-----)
                                 --------|--------|--------|--------|
                                    25       30       35       40
```

 | **4. For Processes, use Confidence Intervals only if Control Charts show that the Process is under control.**

Outliers can have a big effect on Confidence Intervals. So, for Processes, use Control Charts to ensure the Process is stable before doing any test dependent on Confidence Intervals. A Process that is not under control is being affected by Special Causes of Variation outside the Process. These must be addressed before the Process can be accurately analyzed – and improved, if necessary

Related Articles in This book: *Confidence Intervals – Part 1: General Concepts; Alpha, α; Critical Value; Test Statistic; Normal Distribution; Sampling Distribution; z; t, the Test Statistic and Its Distributions; t-tests – Part 1; Proportion*

CONTROL CHARTS – PART 1 (OF 2): GENERAL CONCEPTS AND PRINCIPLES

Summary of Keys to Understanding

1. **All Processes have Variation. A Process can be said to be "under control,"** "stable," or "predictable" **if the Variation is**
   - **confined within a defined range,** and is
   - **random** (shows no pattern)

2. **Such Variation is called <u>Common Cause Variation</u>; it is random "<u>noise</u>" within an under-control Process.**
   Variation which is not Common Cause is called <u>Special Cause Variation</u>. It is a <u>signal</u> that Factors outside the Process are affecting it.

3. **Any Special Cause Variation must be eliminated before one can attempt to narrow the range of Common Cause Variation.**

4. **Control Charts show Variation in a time sequence from left to right.**
   <u>Control Charts</u> tell us whether the Variation is within the prescribed range for Common Cause Variation. The Upper Control Limit (**UCL**) and the Lower Control Limit (**LCL**) define this range.
   <u>Run Rules</u> define whether the Variation is random or whether there is a pattern (e.g., a trend, shift, or cycle). **These patterns are also visible on a Control Chart.**

5. **Data are collected in a number of Rational Subgroups of size $n \geq 4$ under essentially the same conditions.**

Explanation

> **1. All Processes have Variation. A Process can be said to be "under control," "stable," or "predictable" if the Variation is**
>   - **confined within a defined range,** and is
>   - **random** (shows no pattern)

> **2. Such Variation is called** <u>Common Cause Variation</u>**; it is like random "noise" within an under-control Process.**
>   **Variation which is not Common Cause is called** <u>Special Cause Variation</u>**. It is a signal that Factors outside the Process are affecting it.**

In much of statistics, the universe of data is from a Population. For Statistical Process Control – the discipline which uses Control Charts – the universe is a specific Process. There are any number of measurements or Counts that can be taken to monitor the performance of a Process or parts of a Process, e.g., timings, measurements, quantities, defects, costs. Almost none of these will remain absolutely constant over time; there will be Variation.

We can't eliminate Variation from a Process, but we want to have Variation that is predictable. If the Variation is random, and if it is within a defined range, we can be comfortable that the Process itself is consistent. And we can be comfortable in predicting that it will continue to vary in a random fashion within that range. As we'll explain later, the range is usually defined as plus or minus three Standard Deviations.

Predictability is very important in Process management as well as for Customer Satisfaction. If we don't limit Variation to a predictable range, we can't plan – our expenses, our revenues, our purchases, our hours, and so on. And a customer who buys a quarter-pound hamburger may not notice if they get 0.24 or 0.26 pounds of meat. But, they will probably be dissatisfied with 0.18 pounds even if the average has been 0.25. Customers feel Variation, they do not feel averages.

There are two types of Variation: Common Cause and Special Cause.

	Common Cause Variation	Special Cause Variation
	expected	unexpected
	like random "noise"	a "signal" that something is different or has changed
	is part of the Process	outside the Process
	occurs within consistent range of values and no pattern	some data points occur outside that range, or there is a pattern
Example	A manufacturing process step took 21, 19, 18, 20, 22, 20, 18, 19, 22, and 20 seconds.	Sometimes, it takes more than 26 seconds.

> **3. Any Special Cause Variation must be eliminated before one can attempt to narrow the range of Common Cause Variation.**

Here are some examples of Special Causes of Variation:

- an equipment malfunction causes occasional spikes in the size of holes drilled
- an out-of-stock condition causes a customer order to be delayed
- vibration from a passing train causes a chemical reaction to speed up
- a temporarily opened window causes the temperature to drop
- an untrained employee temporarily fills in

Until we eliminate Special Cause Variation, we don't have a Process that we can improve. There are Factors outside the Process which affect it, and that changes the actual Process that is happening in ways that we don't know.

Once we know that we have Special Cause Variation, we can use various Root Cause Analysis methods to identify the Special Cause, so that we can eliminate it. Only then can we use process/quality improvement methods like Lean Six Sigma to try to reduce the Common Cause Variation.

But how do we know if we have Special Cause Variation?

> **4. Control Charts show Variation in a time sequence from left to right.**
>
> **Control Charts tell us whether the Variation is within the prescribed range for Common Cause Variation.** The Upper Control Limit (**UCL**) and the Lower Control Limit (**LCL**) define this range

> **Run Rules** define whether the Variation is random or whether there is a pattern (e.g., a trend, shift or cycle). **These patterns are also visible on a Control Chart.**

Control charts are used to analyze Process Variation over time. So, the data are collected and recorded in time sequence (data point 1 happened first, data point 2 happened next). The results of the Control Chart analysis are displayed in time sequence from left to right.

We said earlier that there are two criteria which determine whether a Process is under control – a defined range and randomness. **Control Charts** display the range and **identify any points outside the range.** Randomness is defined as lack of specific patterns which are non-random. These **patterns are defined by Run Rules** which we will show later. Most softwares which produce Control Charts will also identify any patterns defined by Run Rules.

There are a number of different types of Control Charts. These are described in the article *Control Charts – Part 2: Which to Use When.*

The defined range for Common Cause Variation is generally three Standard Deviations above and below the Center Line (e.g., the Mean). The upper and lower boundaries of this range are called the **Upper Control Limit (UCL)** and the **Lower Control Limit (LCL).** These limits are usually depicted **as colored or dotted horizontal lines; the Center Line is solid. If zero is the lowest possible value** (e.g., when we're measuring time) **a Lower Control Limit calculated to be less than zero is not shown.**

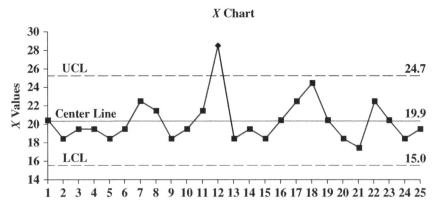

This chart plots the time in seconds that it took to complete a specific step in a manufacturing Process. All the times except for the Outlier at 28 seconds are between the Lower Control Limit (LCL) and the Upper Control Limit (UCL). So, they are due to Common Cause Variation. But because of the one value outside the range, this Process is <u>not</u> under control. We need to take a look at what may be the Special Cause of that anomaly and try to fix it.

Control Limits are not the same as Specification Limits. Control Limits are statistically calculated from the data; they are used to help make a Process stable and predictable. Specification limits are usually set by a management decision with customer satisfaction in mind.

<u>Run Rules</u> define patterns which indicate the Variation is not random.

There are a number of Run Rules, and some patterns are not always easy to spot by eyeballing charts. Fortunately, the same software which produces Control Charts will usually also identify patterns described by the Run Rules. Here are some common patterns which indicate non-random (Special Cause) Variation. A Sigma is a Standard Deviation.

- <u>Trend</u>: 6 consecutively increasing or 6 consecutively decreasing points
- <u>Shift</u> in the Mean: 8 consecutive points on the same side of the Center Line
- <u>Cycle</u>: 14 consecutive points alternating up and down
- 2 out of 3 points beyond 2 Sigma and on the same side of the Center Line
- 4 out of 5 points beyond 1 Sigma and on the same side of the Center Line
- 15 consecutive points within 1 Sigma of the Center line

---

 5. Data are collected in a number of Rational Subgroups of size $n \geq 4$ under essentially the same conditions.

Statistical Process Control (SPC) is different from Inferential Statistics (e.g., *t*-tests or ANOVA). In Inferential Statistics, we collect a Sample of data (the larger, the better – up to a point). And we're trying to infer the value of a Parameter (e.g., the Mean) in the Population or Process from the value of the corresponding Statistic in the Sample.

In SPC, we collect a number of small Samples in order to identify Variation over time. The Samples could be as small as 4 or 5, and **we collect 25 or more of them**. These are a special kind of Sample called the Rational Subgroup.

Rational Subgroups are comprised of individual data points collected from a Process under the same conditions. For example,

- the same step in the Process
- the same operator
- the same machine, etc.
- close together in time

These conditions are the same, because we want to eliminate (block) them as potential causes of Variation.

With the exception of the X ("Individuals") chart, like the one shown above, Rational Subgroups are composed of more than one data point. So we can calculate a Statistic (e.g., the Mean, Proportion, Standard Deviation, or Range) for each Rational Subgroup. **It is the value of that Statistic which gets plotted in the Control Chart.**

The Part 2 article will describe various different Control Charts and tell you which to use when.

Related Articles in This Book: *Control Charts – Part 2: Which to Use When*; *Variation/Variability/Dispersion/Spread*

CONTROL CHARTS – PART 2 (OF 2): WHICH TO USE WHEN

This article assumes knowledge of the content of the Part 1 article.

Summary of Keys to Understanding

1. Continuous/Measurement Data

Statistic plotted	Sample Size (*n*)	Distribution	Control Chart(s) to Use
Standard Deviation and Mean	3+	–	*s* and *X*-bar
Range and Mean	2	–	*R* and *X*-bar
Moving Range	1	Normal	MR and *X*

Recommendation:

- **Collect data in 25 or more Rational Subgroups** (see the Part 1 article) **with** $n \geq 4$.
  - **Use the** *s* **chart first to determine if the Variation is under control.** If not, no need to bother with the *X*-bar chart.
- **Use the** *X*-bar **chart to determine if the Central Tendency is under control.**
- **Use the** *X* (aka **Individuals** aka **I**) **Chart to test for non-random Run Rule patterns.**

2. Discrete/Count data

Statistic plotted	Sample Size (*n*)	Counted Item	Control Chart(s) to Use
Proportion	All the same	units	np
Proportion	Varies		p
Count	All the same	occurrences	c
Occurrences	Varies		u

---

3. **Run Rules:** non-random patterns of Variation can be found with the following charts
   - *X*-bar
   - *X* (if the data are Normally distributed)
   - np
   - *c*

Explanation

There are many different types of Control Charts; this article describes some of the most common.

The 1st criterion for deciding what Control Chart to use is the type of data.

Much as there are different Distributions **for Continuous vs. Discrete data, there are different Control Charts.** Continuous data come from Measurements of things like temperature, length, and weight. Discrete data come from integer Counts of things like defects.

Continuous data Control Charts are sometimes call Shewhart Control Charts. Discrete data Control Charts are sometime called Attributes Control Charts, because they count the number of times an Attribute (say, a defect) is present.

1. Continuous Data

For Continuous data, we want both the Variation (Standard Deviation or Range) **and the Central Tendency** (Mean, Mode, or Median) **of our Process to be under control. So, we should look at a Control Chart for each.** That is why software often produces them in pairs. (s and X-bar, R and X-bar, MR and X).

Recommendation for Continuous Data

- **Collect data in 25 or more Rational Subgroups** (see the Part 1 article) **with $n \geq 4$.**
- **Use the s chart first to determine if the Variation is under control.**
  - If not, no need to bother with the X-bar chart.
- **Use the X-bar chart to determine if the Central Tendency is under control.**
- **Use the X** (aka **Individuals** aka **I**) **Chart to test for non-random Run Rule patterns** (if the software for the above charts doesn't automatically provide that information) and also to identify individual out-of-control points.

Control Charts	Statistics plotted	Rational Subgroup Size (n)
s and X-bar	Standard Deviation and Mean	4+

If we have sufficient time, money, and opportunity, the best approach is to use the Rational Subgroup (see the Part 1 article) type of Sample. Each

Subgroup has the same Size. The minimum Size is $n = 4$, although more is better. In a Rational Subgroup, the individual data points are collected close together in time, under essentially the same conditions. **We need at least 25 such Rational Subgroups to show Variation over time.**

Software usually shows these two Control Charts together, with the *X*-bar (aka Averages) Chart on top and the *s* Chart below. Unfortunately, this can lead the user to consider the *X*-bar first. The problem is that Control Limits (UCL and LCL) shown on the *X*-bar Chart are calculated using a Statistic which measures Variation (the Range). **If the Variation is out of control, the Control Limits on the *X*-bar charts will be meaningless. So, examine the *s* Chart first. If it is out of control, the Process is out of control.**

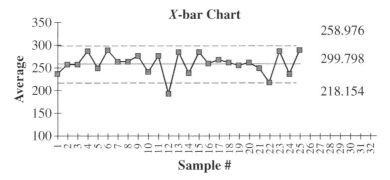

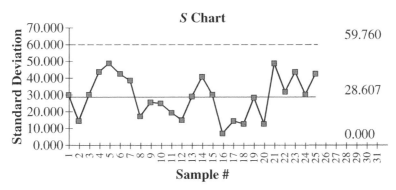

Other Continuous Data Charts

Sometimes, cost or the need to use historical data prohibits us from using Rational Subgroups of Size $n \geq 4$. The results won't be as accurate, but we can use smaller Sample Sizes and the Range instead of the Standard Deviation to determine if the Variation is under control. **Examine the *R* (Range) Chart first. If it is out of control, the Process is out of control, and the *X*-bar Chart is not usable.**

As we explain in the article *Variation/Variability/Dispersion/Spread*, Range is an inferior measure of Variation compared with Standard Deviation. For Sample Size $n = 2$, there is not much difference between the two. But as n increases, s provides more and more accuracy. By the time $n > 9$, the difference in accuracy is generally deemed to be intolerable. That is why **the R Chart <u>can</u> be used for Sample Sizes $n = 2$ to 9. But, for any Size larger than 2, it's better to use the s Chart.**

Statistic plotted	Sample Size (n)	Distribution	Control Chart(s) to Use
Range and Mean	2	–	R and X-bar
Moving Range	1	Normal	MR and X

If $n = 1$, we can't calculate any kind of Statistic <u>within</u> each Sample. But we <u>can</u> calculate a Range for two adjacent data points. This is called a **Moving Range (MR).**

Often paired with the MR is the **X Chart** (aka the **Individuals** or the **I Chart**) which plots the individual data points – the x's. If the data are Normal, we don't need to use Means and the Central Limit Theorem in order to know that 99.7% of the data should lie within three Standard Deviations (Sigmas) of the Mean. We can just use the raw data, as in the X Chart.

The X Chart may seem less sophisticated than the X-bar or s Charts, but it **has the advantage of showing us the raw data.** This enables us to see individual points beyond the Control Limits, **which can be helpful in investigating the causes of Variation.** So, many practitioners always include it in their assessments.

If the data are not Normal, and the Sample Size is 1 we cannot calculate 3 Sigma Upper and Lower Control Limits, so **we don't have a Control Chart.** What's left is an X chart without Control Limits. This is called a Run Chart. It's not a Control Chart, but it can be used for checking at least a few of the Run Rules (see the Part 1 article) which identify non-random patterns of Variation. These patterns also indicate an out-of-control Process.

2. Discrete Data

Counted Item	Sample Size (n)	Control Chart to Use	Run Rules?
Units	All the same	np	applicable
	Varies	p	no
Occurrences	All the same	c	applicable
	Varies	u	no

Discrete data are integers, such as Counts. Counts are non-negative. In contrast to Continuous data, there are no intermediate values between consecutive integer values.

For Discrete data, the first criterion in selecting a Control Chart is type of thing counted – Units or Occurrences.

Let's say we're counting defects in a Process which manufactures shirts. The shirt is the Unit; the number of defects is the number of Occurrences.

	number of Units with defects	number of Occurrences
1 shirt with 3 defects	1	3

FYI: the reason there are two different sets of Discrete data Control Charts is that they are based on two different Discrete Distributions – the Binomial Distribution (used for Units) and the Poisson Distribution (used for Occurrences).

For Units – the np and p Charts – the Statistic plotted is the Proportion of Units (with defects). **If the Samples are all the same Size, use the np chart; otherwise, use the p Chart.**

For Occurrences, use the c (Count) Chart if the Samples are all the same Size; otherwise use the u chart.

---

3. **Run Rules**: non-random patterns of Variation can be found with the following charts
   - X-bar
   - X (if the data are Normally distributed)
   - np
   - c

See the Part 1 article for a list of common Run Rules.

Related Articles in This Book: *Control Charts – Part 1: General Concepts & Principles*; *Variation/Variability/Dispersion/Spread*; *Standard Deviation*; *Proportion*; *Binomial Distribution*; *Poisson Distribution*

CORRELATION – PART 1 (OF 2)

Summary of Keys to Understanding

 1. Correlation is observed when two Variables either increase together or decrease together in a roughly linear pattern.

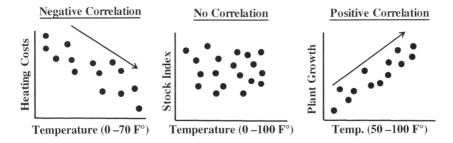

Negative Correlation	No Correlation	Positive Correlation

Heating Costs — Temperature (0 –70 F°) — Stock Index — Temperature (0 –100 F°) — Plant Growth — Temp. (50 –100 F°)

 2. Covariance is a Statistic or a Parameter which **can tell us the direction of a Correlation** between two paired Variables, x and y, **from data consisting of (x, y) pairs.**

 3. Covariance can be conceptually viewed as a 2-dimensional Variance of the (x, y) data points about the point with their average values, $(\bar{x}, \bar{y})$.

$$\text{Sample: } \mathbf{Cov}(x, y) = s_{xy}\frac{\sum(x - \bar{x})(y - \bar{y})}{n - 1}$$

$$\text{Population: } \mathbf{Cov}(x, y) = \sigma_{xy} = \frac{\sum(x - \mu_x)(y - \mu_y)}{N}$$

 4. Covariance can <u>not</u> tell us the strength of the Correlation.

 5. When normalized or standardized, the Covariance becomes a Correlation Coefficient, which is a very useful measure of the direction and strength of the Correlation. That is the subject of the next article, *Correlation – Part 2.*

Explanation

> **1. Correlation is observed when two Variables either increase together or decrease together in a roughly linear pattern.**

Correlation is Negative when larger values for one Variable are paired with smaller numbers of the other. Positive Correlation is the opposite – the values of both Variables grow together.

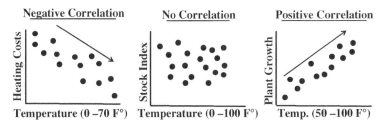

> **2. Covariance** is a Statistic or a Parameter which **can tell us the direction of a Correlation** between two paired Variables, x and y, **from data consisting of (x, y) pairs.**

The Covariance calculated from a Sample of data is, by definition, a Statistic. The Covariance calculated from all the data in a Population or Process, is, by definition a Parameter.

Covariance is calculated for two paired Variables, which we will label x and y. It is a signed number or zero. The sign indicates the direction of the Correlation.

What do we Mean by "paired" variables?

Each data point has 2 values, 1 for the x Variable and 1 for the y Variable. Each data point can be written as (x, y). For example, if we are comparing height (say, x) and weight (y), each (x, y) data point is from one person; it contains a height from a person and that same person's weight. You cannot calculate a Covariance between height and weight by using the height of one person and the weight of another.

It is important to note that **Correlation does not attempt to demonstrate Cause and Effect.** That is the purpose of Regression analysis, which can be considered to be an extension of Correlation analysis.

So, although we use the terms x and y for the paired variables, **the existence of Covariance does not mean that $y = f(x)$.**

> **3. Covariance can be conceptually viewed as a 2-dimensional Variance of the (x, y) data points about the point with their average values, $(\bar{x}, \bar{y})$.**

For a single Variable, x, Variance is a measure of Variation of the values of x in the data about their Mean, (symbol $\bar{x}$ for a Sample, or μ for a Population or Process). Covariance is a measure of Variation of the values of 2-Variable data points (x, y)'s about the point made up of the Means of x and y – the point $(\bar{x}, \bar{y})$. So, we can think of Covariance as a 2-Variable counterpart to the Variance.

Variance (1 Variable) Formulas

$$\text{Sample: } s^2 = \frac{\sum(x - \bar{x})^2}{n - 1} \qquad \text{Population: } \sigma^2 = \frac{\sum(x - \mu_x)^2}{N}$$

where n and N are the Sample Size and Population Size, respectively

Covariance (2 Variable) Formulas

$$\textbf{Sample}: \text{Cov}(x, y) = s_{xy} = \frac{\sum(x - \bar{x})(y - \bar{y})}{n - 1}$$

$$\textbf{Population}: \text{Cov}(x, y) = \sigma_{xy} = \frac{\sum(x - \mu_x)(y - \mu_y)}{N}$$

Here's an example; in the United States, we measured height and weight for 10 individuals.

Covariance of Height (inches) and Weight (pounds)					
Individual	Height(x)	Weight(y)	x-Mean(x)	y-Mean(y)	Product
#1	70	180	2.3	21	48.3
#2	65	125	−2.7	−34	91.8
#3	67	140	−0.7	−19	13.3
#4	71	195	3.3	36	118.8
#5	62	105	−5.7	−54	307.8
#6	73	210	5.3	51	270.3
#7	68	190	0.3	31	9.3
#8	65	110	−2.7	−49	132.3
#9	70	200	2.3	41	94.3
#10	66	135	−1.7	−24	40.8
Total	677	1590			
Means	67.7	159.0	Sum of Products:		1127.0
Divide the Sum by $n - 1 = 9$ to get the Covariance: 125.2 inch-pounds					

Note that we divided the sum by $n - 1 = 9$, because this is a Sample.

You may be wondering, "What's an inch-pound?" Good question for which there is no good answer. This is one reason why the Covariance is of limited usefulness.

Now, let's say after being measured and weighed as above, each of the 10 subjects walked across the hall and were weighed and measured again – this time by some researchers visiting from Europe. Below are the data they recorded:

Covariance of Height (meters) and Weight (kilograms)						
Individual	Height (x)	Weight (y)		x-Mean(x)	y-Mean(y)	Product
#1	1.8	81.7		0.1	9.5	0.6
#2	1.7	56.8		−0.1	−15.4	1.1
#3	1.7	63.6		0.0	−8.6	0.2
#4	1.8	88.5		0.1	16.3	1.4
#5	1.6	47.7		−0.1	−24.5	3.5
#6	1.9	95.3		0.1	23.2	3.1
#7	1.7	86.3		0.0	14.1	0.1
#8	1.7	49.9		−0.1	−22.2	1.5
#9	1.8	90.8		0.1	18.6	1.1
#10	1.7	61.3		0.0	−10.9	0.5
Total	17.2	721.9				
Mean	1.72	72.2		Sum of Product:		13.0
Divide the Sum by $n - 1 = 9$ to get the **Covariance: 1.4 meter-kilograms**						

You can see the difficulty in using Covariance. Not only do we have meaningless units, we have widely varying values, 125.2 and 1.4, for the same data.

 | **4. Covariance _cannot_ tell us the strength of the Correlation.**

One thing we can say from both sets of measurements above is that there is a positive Correlation. That is, as height increases, weight also increases. So, we can use the sign of these numbers (positive) to tell us the direction of Correlation (positive).

But how good is this Correlation? How strong is it? We can't use the values of the numbers, because the units are meaningless and we would have to make an arbitrary choice between whether the strength was 125.2 or 1.4.

So the numerical values of the Covariance are not used. We only use the <u>sign</u> – positive or negative – of the Covariance to tell us the direction of the correlation.

We can see that **Covariance itself is of limited use** because

- It can tell us the direction of Correlation, but not the strength.
- It is in units that are often meaningless, e.g., meter-kilograms.

So, it is easy to see that a better measure of Correlation is needed.

> 5. **When normalized or standardized, the Covariance becomes the Correlation Coefficient, a measure of the direction and strength of the Correlation**.

The primary purpose of Covariance is to serve as an interim step in the calculation of the Correlation Coefficient – which is the subject of the next article, *Correlation Part 2*.

Related Articles in This Book: *Correlation Part 2*; *Variance*

CORRELATION – PART 2 (OF 2)

Summary of Keys to Understanding

 1. Always plot the data first. Statistics alone can be misleading.

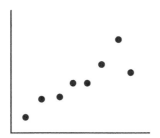 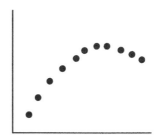

 2. The Correlation Coefficient, r, is a measure of Correlation. It is the "standardization" or "normalization" of Covariance.

$$r = \frac{\text{Cov}(x, y)}{s_x s_y}$$

 3. The Correlation Coefficient, r, ranges from -1 to $+1$. $r = 0$ indicates no Correlation. $r = -1$ and $r = +1$ indicate a perfect negative or positive Correlation, respectively. But perfection almost never happens, and there are different opinions on where to set the thresholds for "strong" or "weak" Correlation.

 4. Correlation is <u>not</u> Causation.

 5. Establishing Correlation is a prerequisite for Linear Regression.

Explanation

> **1. Always plot the data first. Statistics alone can be misleading.**

It has been said that the first three laws of statistics are 1. Plot the data, 2. Plot the data, and 3. Plot the data. The need for this is never more apparent than in Correlation. Scatter Plots like those below are probably the most useful type for this purpose.

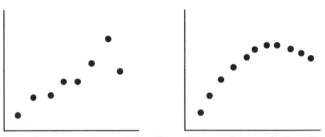

Statistics alone can be misleading. The values of the Correlation Coefficient, r, for these two plots are almost identical – and they both indicate a very strong Linear Correlation.

That makes sense for the one on the left. However the one on the right is not linear at all. That data would more likely to be approximated by a polynomial curve.

Since Correlation analysis is only about linear relationships, **if the two Variables are Correlated, the data should at least roughly cluster about an imaginary slanted line – throughout its entire range. If you don't see that in the plot, it would be unwise to proceed** with trying to use Correlation Analysis to prove differently.

Why a slanted line? A horizontal or vertical line would mean that one Variable can increase or decrease while the other stays the same.

> **2. The Correlation Coefficient, r, is a measure of Correlation. It is the "standardization" or "normalization" of Covariance.**

In Part 1, we learned that Covariance has some serious shortcomings. It is in meaningless units like "kilogram-meters." As a result, it is not useful as a measure of the strength of the relationship between the variables. It can only tell us the direction (negative, zero or positive).

The Covariance is in units which are the product of the units of x and the units of y. To get rid of this confusion, we can divide Covariance by something which has those same units.

Since Standard Deviations are in the units of the data, the product of the Standard Deviation of x and the Standard Deviation of y will be in the units of xy, e.g., kilograms, multiplied by the units of y, e.g., meters that yields "kilogram-meters."

So, **if we divide Covariance by the product of the Standard Deviations of x and y, we can eliminate the units and "standardize" the Covariance**. The following formulas are for Samples. For Populations or Processes, use sigma, σ, instead of s and N instead of $n-1$.

The Correlation Coefficient is defined as: $r = \dfrac{\text{Cov}(x,y)}{s_x s_y}$

r is also known as "**Pearson's r**" or the "Pearson product-moment correlation coefficient" (don't ask). There are other alternatives for the Correlation Coefficient, but this is the most widely used.

We divide the Covariance by the product of the Standard Deviations of x and y to "standardize" the Covariance. As a result, r is a unit-less number.

The following indented section is optional. It shows two other (but equivalent) formulas for r. Looking at r in several different ways may be helpful for some people.

The Covariance can be shown as the Standard Deviation of xy, i.e., $\text{Cov}(x,y) = s_{xy}$. That gives us our **2nd equivalent formula for r**:

$$r = \frac{s_{xy}}{s_x s_y}$$

In the Part 1 article, we said: $\text{Cov}(x,y) = s_{xy} = \dfrac{\sum(x - \bar{x})(y - \bar{y})}{n - 1}$

Using simple algebra, we move things around to get:

$$r = \frac{1}{n - 1} \sum \left(\frac{x - \bar{x}}{s_x} \right) \left(\frac{y - \bar{y}}{s_y} \right)$$

Those terms in the parentheses may look familiar. z is the Test Statistic for the z-Distribution, which is also known as the Standard Normal Distribution.

$z_x = \dfrac{x - \bar{x}}{s_x}$ and likewise for y. If we make those substitutions, we can see that r can be expressed in a **3rd equivalent formula, the "normalized" Covariance**.

$$r = \frac{\sum z_x z_y}{n - 1}$$

where n is the number of (x,y) pairs in our data.

To put the three formulas for the Correlation Coefficient all in one place: Remember, these are just different algebraic manipulations of the same thing.

$$r = \frac{\text{Cov}(x,y)}{s_x s_y} = \frac{s_{xy}}{s_x s_y} = \frac{\sum z_x z_y}{n-1}$$

The above formulas are for Samples, so the terms are all Latin characters. For Populations or Processes, substitute the Greek letter rho (ρ) for r and the Greek letter sigma (σ) for s.

 3. The Correlation Coefficient, r, ranges from -1 to $+1$. $r = 0$ indicates no Correlation. $r = -1$ and $r = +1$ indicate a perfect negative or positive Correlation, respectively. But perfection almost never happens, and there are different opinions on where to set the thresholds for "strong" or "weak" Correlation.

In the Part 1 article, we showed that the value of Covariance is different for different units of measure. The standardization or normalization that we described above eliminates this shortcoming. **The Correlation Coefficient, r, tells us both the direction and the strength of the Correlation.**

Evidence of Correlation	e.g., Less Rigorous Standard	e.g., More Rigorous Standard
very strong	0.7 – 1.0	0.81 – 1.00
strong	0.5 – 0.7	0.61 – 0.80
moderate	0.3 – 0.5	0.41 – 0.60
weak	0.1 – 0.3	0.21 – 0.40
none	0.0 – 0.1	0.00 – 0.20

(Negative Correlation thresholds are the same, only negative). In the social sciences, the phenomena being analyzed are not as precisely governed by the laws of science as are engineering and scientific phenomena or processes. So, the standard in social sciences tends to be less rigorous.

 4. Correlation is <u>not</u> Causation.

Correlation analysis does not attempt to determine Cause and Effect. **The value of r** may tell us that x and y have a strong Correlation, but it **cannot tell us that x causes y or vice versa.** It may well be neither. There could be one or more unknown "lurking" Variables which influence both *x* and *y*.

We can't even use the fact that *r* is near zero to conclude that there is no cause-effect relationship, just that there is no <u>linear</u> cause-effect relationship.

Regression is the tool for trying to explore cause and effect. But even Regression is limited. **Cause and Effect cannot be proven by statistics alone.**

But the results from a Correlation analysis may give us some insights to test further with subject-matter-expert knowledge of the real-world interactions of the Factors involved and Linear Regression, followed by Designed Experiments.

 | **5. Establishing Correlation is a prerequisite for Linear Regression.**

You can't use Linear Regression unless there is a Linear Correlation. The following compare-and-contrast table may help in understanding both concepts.

	Correlation	**Linear Regression**
Purpose	Description, Inferential Statistics	Prediction, Designed Experiments
Statistic	r	r, R^2, R^2-adjusted
Variables	Paired	2, 3, or more
Variables	No differentiation between the Variables.	Dependent y, Independent x, $y = f(x)$
Fits a straight line through the data	Implicitly	Explicitly: $y = a + bx$
Cause and Effect	Does not address	Attempts to show

Correlation analysis <u>describes</u> the present or past situation. It uses Sample data to <u>infer</u> a property of the source Population or Process. There is no looking into the future. The purpose of Linear Regression, on the other hand, is to define a Model (a linear equation) which can be used to <u>predict</u> the results of Designed Experiments.

Correlation mainly uses the Correlation Coefficient, *r*. Regression also uses *r*, but employs a variety of other Statistics.

Correlation analysis and Linear Regression both attempt to discern whether two Variables vary in synch. Linear Correlation is limited to two Variables, which can be plotted on a 2-dimensional x–y graph. Linear Regression can go to three or more Variables/dimensions.

In Correlation, we ask to what degree the plotted data forms a shape that seems to follow an imaginary line that would go through it. But we don't try to specify that line. In Linear Regression, that line is the whole point. We calculate a best-fit line through the data: $y = a + bx$.

Correlation Analysis does not attempt to identify a Cause-Effect relationship, Regression does.

Related Articles in This Book: *Correlation – Part 1*; *Regression – Parts 2–4*; *Standard Deviation*; *z*

CRITICAL VALUE

It may be helpful to read the article "Alpha, α" before reading this article.

Summary of Keys to Understanding

 1. A Critical Value is derived from the Significance Level, α, and the Probability Distribution of a Test Statistic (like z, t, F, or Chi-square).

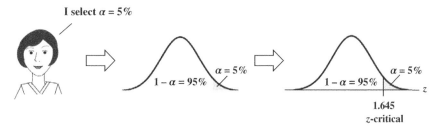

 2. In Inferential Statistics, comparing a Test Statistic (e.g., z) **to its Critical Value** (z-critical) **is statistically the same as comparing the p-value to Alpha.**

Areas under the curve (right tail) α: ▭ p: ▨	z-critical z	z z-critical
	$p \leq \alpha$ $z \geq z$-critical	$p > \alpha$ $z < z$-critical
The observed difference, change, or effect is:	Statistically Significant	not Statistically Significant

 3. Critical Values are used in defining the boundaries of Confidence Intervals.

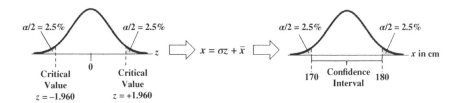

Explanation
It may be helpful to read the article "Alpha, α" before reading this article.

 1. A Critical Value is derived from the Significance Level, α, and the Probability Distribution of a Test Statistic (like z, t, F, or Chi-square).

In Inferential Statistics, we estimate the value of a Population or Process Parameter, say the Mean, from the corresponding Statistic in a Sample. Being an estimate, there is a Probability of error. Before collecting the Sample data, we select a tolerable level of Alpha (False Positive) Error for this estimate. If we want a 95% Confidence Level of avoiding an Alpha Error, **we select a Significance Level, Alpha** (denoted by $α$), of 5%.

I want to be 95% confident of avoiding an Alpha Error.

So, I'll select $α = 5\%$.

A Test Statistic (e.g., z, t, F and Chi-square) **is one that has a Probability Distribution**. That is, for any given value of the Test Statistic, the Probability of that value occurring is known. Graphically, it is the height of the Distribution's curve above that value. These Probabilities are available in tables or via software.

The total of the Probabilities of a range of values is called a Cumulative Probability. The Cumulative Probability of all values under the curve is 100%. Alpha is a Cumulative Probability. In the following illustration of a 1-sided test, we shade the rightmost 5% of the area under the curve of the z-Distribution to represent Alpha.

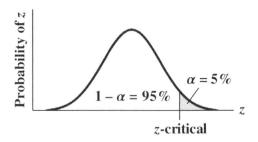

We then draw a vertical line to mark the boundary between the α and $1 - \alpha$ areas under the curve. The intersection of this line with the horizontal axis gives us a number which is the Critical Value for a given value of Alpha. Given a Test Statistic Distribution and a value for Alpha, tables or software can provide us the Critical Value.

In this example, selecting $\alpha = 5\%$ and using the z-Distribution results in 1.645 as the Critical Value of z, also known as "z-critical."

The Cumulative Probability of z-critical occurring, and that of all z-scores more extreme (farther from The Mean) than z-critical, is Alpha. When looking up or calculating Critical Values, we also need to know whether our test is left-tailed, right-tailed, or 2-tailed.

1-Sided, Right-Tailed

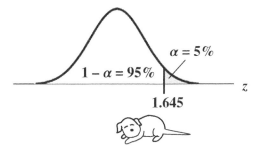

The above example is a **1-sided** or **1-tailed** test, specifically, a **right-tailed** test. You may find tables that are right-tailed or left-tailed (or 2-tailed). They will often show a graph like the one above (without the dog) to help identify which one it is.

1-Sided, Left-Tailed

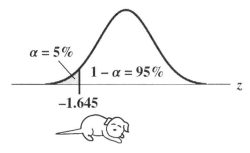

Note that **the Critical Value for z in a left-tailed test is just the negative of the Critical Value for the right-tailed test.** So, if the Distribution is left–right symmetrical around zero (like z and t), you can use a right-tailed table to get a left-tailed value. **You cannot do this for Test Statistics like F and Chi-square, which have asymmetrical (skewed) Distributions. Their left and right Critical Values must be calculated separately.**

2-Sided (aka 2-Tailed)

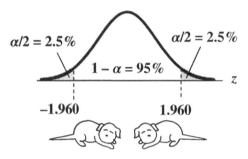

For 2-sided tests, use the Critical Value for $\alpha/2$ for each of the two shaded areas. This results in Critical Values more extreme (farther away from the Mean) than the single value for α (e.g., 1.960 vs. 1.645 for z, with $\alpha = 5\%$).

 2. In Inferential Statistics, comparing a Test Statistic (e.g., z) **to its Critical Value** (e.g., z-critical) **is statistically the same as comparing the p-value to Alpha.**

Let's say we calculate $z = 1.2$ from our Sample data. What is the Probability of any given Sample of the same size getting that value or larger? We can use the z-Distribution to give us that Probability, 11.5% in this example. We find **the area under the curve for z and all values more extreme** (farther from the Mean). **That area is the p-value (p).**

So, **like α and z-critical, p and z convey the same information.** Using the Distribution, we can get either one from the other – p from z or z from p.

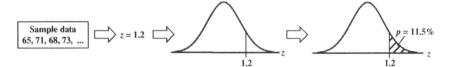

The p-value (p) is a Sample's actual Probability of an Alpha Error. We compare this with Alpha (α) – the maximum tolerable Probability of Alpha Error which we selected.

Comparing a Test Statistic value to a Critical Value (e.g., z to z-critical) **gives the same result as comparing p to α. Either of these comparisons will tell us the conclusion to be drawn from the analysis.**

Areas under the curve (right tail) α: ▭ p: ▱	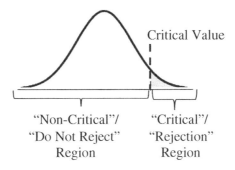 z-critical	z-critical
	$p \le \alpha$ $z \ge z$-critical	$p > \alpha$ $z < z$-critical
The observation from the Sample data is an accurate estimate for the Population or Process as a whole.	True	False
Null Hypothesis	Reject	Accept (Fail to Reject)
The observed difference, change, or effect is:	Statistically Significant	not Statistically Significant

To further your understanding of how all this works, it may be helpful to read the article *Alpha, p, Critical Value and Test Statistic – How They Work Together*. Also, if you're having trouble remembering which way the inequality signs go, see the article *p, t, and F: ">" or "<" ?*

The terms Critical Region, Rejection Region, etc. are sometimes used to describe the shaded and unshaded areas, as shown below.

Critical Value

"Non-Critical"/   "Critical"/
"Do Not Reject"   "Rejection"
Region   Region

3. **Critical Values are used in defining the boundaries of Confidence Intervals.**

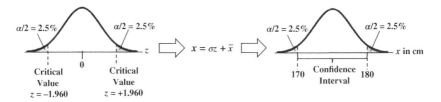

$\alpha/2 = 2.5\%$ $\alpha/2 = 2.5\%$ $\alpha/2 = 2.5\%$ $\alpha/2 = 2.5\%$

$x = \sigma z + \bar{x}$

Critical   Critical   170   Confidence   180
Value   Value   Interval
$z = -1.960$ $z = +1.960$ x in cm

Critical Values are in units of the Test Statistic, like z's. We can convert Critical Values into the real-world units of the Sample data, using formulas like the one shown above. This converts the Critical Values into Confidence Limits, which define a Confidence Interval.

The Confidence Intervals method is one of the two main methods of Inferential Statistics; Hypothesis Testing is the other.

Related Articles in This Book: *Alpha,* α; *Distributions – Part 1: What They Are*; *Inferential Statistics*; *Test Statistic*; z; *Alpha, p, Critical Value and Test Statistic – How They Work Together*; *p, p-Value*; *p, t, and F: ">" or "<" ?*; *Confidence Intervals*; *Hypothesis Testing – Parts 1 and 2*; *Reject the Null Hypothesis*; *Fail to Reject the Null Hypothesis*

DEGREES OF FREEDOM

Summary of Keys to Understanding

 1. **Degrees of Freedom** (symbol: df or the Greek letter v) **is a way of adjusting for the additional error introduced when one Statistic is used to calculate another.**

$$\text{Sample Mean} = \frac{\sum x}{n} \qquad \text{Sample Variance} = \frac{\sum (x - \text{Sample Mean})}{n - 1}$$

 2. **The effect of the adjustment gets smaller for larger Sample Sizes.**

 3. **The individual members of some families of Distributions are specified by their df.**

 4. **Formulas for Degrees of Freedom vary by the Statistics and the test they are used in**

Statistic	df	Explanation
ANOVA: Mean Sum of Squares Within (MSW)	$N - k$	N: total # of all data points
ANOVA: Mean Sum of Squares Between (MSB)	$k - 1$	k: # of groups
χ^2	$n - 1$	n: Sample Size
χ^2 test for Goodness of Fit	$n - 1$	n: # of categories
χ^2 test for Independence	$(r - 1)(c - 1)$	# of rows and columns
χ^2 test for Variance	$n - 1$	n: Sample Size
F	$n_1 - 1$ and $n_2 - 1$	n_1 and n_2: Sizes of the 2 Samples
t	$n - 1$	n: Sample Size
1-Sample t-test, and Paired t-test	$n - 1$	
2 (Independent)-Sample t-test	$n_1 + n_2 - 2$	n_1 and n_2: Sizes of the 2 Samples

Statistics from A to Z: Confusing Concepts Clarified, First Edition. Andrew A. Jawlik.
© 2016 John Wiley & Sons, Inc. Published 2016 by John Wiley & Sons, Inc.

Explanation

> **1. Degrees of Freedom** (symbol: df or the Greek letter v) **is a way of adjusting for the additional error introduced when one Statistic is used to calculate another.**

Notation: Degrees of Freedom is often abbreviated as "df" in text; the Greek letter nu (v) is commonly used in formulas or equations.

A Statistic is a numerical property of a Sample, for example, the Sample Mean or Sample Variance. **A Statistic is an estimate** of the corresponding property ("Parameter") in the Population or Process from which the Sample was drawn. Being an estimate, it will likely not have the exact same value as its corresponding population Parameter. The difference is **the error in the estimation**.

So, **if we calculate a Statistic entirely from data values, there is a certain amount of error.**

Example: Sample Mean

The Sample Mean is calculated entirely from the values of the Sample data. It is the sum of all the data values in the Sample divided by the number, n, of items in the Sample. **There is one source of error in its formula – the fact that it is an estimate** because it does not use all the data in the Population or Process.

$$\text{Sample Mean} = \frac{\sum x}{n}$$

If we then use that Statistic to calculate another Statistic, it brings its own estimation error into the calculation of the second Statistic. This error is in addition to the second Statistic's estimation error. This happens in the case of the Sample Variance.

Example: Sample Variance

$$\text{Numerator for Sample Variance:} = \sum (x - \text{Sample Mean})^2$$

The numerator of the formula for Sample Variance includes the Sample Mean. It takes each data value (the x's) in the Sample and subtracts from it the Sample Mean. Then it squares and sums all those subtracted values.

So, **the Sample Variance has two sources of error:**

– it is an estimate from Sample data

– the estimation error from the Sample Mean

It would be good to somehow distinguish between the effects of the one source of error for the Mean and the two sources of error for the Variance. **The Degrees of Freedom is intended to adjust for the additional error introduced when one Statistic is used to calculate another.**

We don't need to make this adjustment for the Sample Mean, but we do need to do so for the Sample Variance. We divide by $n-1$, instead of n.

$$\textbf{Sample Mean} = \frac{\sum x}{n} \qquad \textbf{Sample Variance} = \frac{\sum (x - \textbf{Sample Mean})}{n-1}$$

(The following is an alternative description of the concept of Degrees of Freedom. Feel free to skip this if you're comfortable with your understanding at this point.)

Another way that Degrees of Freedom is described is "The number of independent pieces of information that go into the calculation of a Statistic." To illustrate, let's say we have a Sample of $n = 5$ data values: 2, 4, 6, 8, and 10.

When we calculate the Sample Mean, we have 5 independent pieces of information – the 5 values of the data. They are independent because none of the values are dependent on the values of another. So, for the Mean, df = 5.

$$\text{Sample Mean} = (2 + 4 + 6 + 8 + 10)/5 = 30/5 = 6$$

But, when we calculate the Sample Variance, we use the Mean as well as the 5 data values. The Mean is not an independent piece of information, because is it dependent on the other 5 values.

Also, when we include the Mean, we only have 4 independent pieces of information left. If we know that the Mean is 30, and we have the data values 2, 4, 6, and 8, then we can calculate that the last data value has to be 10. So, 10 no longer brings independent information to the table.

 | **2. The effect of the error adjustment gets smaller for larger Sample Sizes.**

n	$n-1$	% adjustment
10	9	10%
100	99	1%

The adjustment is applied because using Sample data will never be as accurate as using all the Population or Process data would be (if that were possible and practical). However, using a larger Sample will always have less inherent error than using a smaller Sample.

For large values of n, the difference between n and $n - 1$ can be negligible.

> ### 3. The individual members of some families of Distributions are specified by their df.

The z-Test Statistic has only one Probability Distribution – the Standard Normal Distribution. But the Probabilities for t, F, and Chi-Square (χ^2) are described by families of Distributions. The individual members of these families are defined by their Degrees of Freedom.

For example, pictured below are three members of the Chi-Square family of Distributions. The shaded area represents Alpha = 5%. And the numbers below the horizontal axis represent the Critical Value of χ^2 for $\alpha = 5\%$.

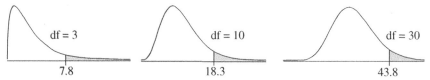

For small values of Degrees of Freedom, the bulk of the χ^2 Distribution is on the left, close to zero, with the tail skewed to the right. As df increases, the bulk moves to the right and the shape becomes more symmetrical. And the Critical Value, which is the boundary of the 5% shaded area under the curve, is a larger number.

The F Test Statistic is the ratio of the Variances of two Samples. The Sample Sizes, n_1 and n_2 can be different. The two Degrees of Freedom from the two Samples are required to specify an individual Distribution. The shapes of these Distributions are similar to those of Chi-Square, and they go through similar changes as the Degrees of Freedom increase.

The shape of the Distribution of the t Test Statistic is the same as the familiar bell-shaped curve of the Normal Distribution. For smaller values of df, the t-Distribution is wider than the Normal. As df increases, the shape narrows and the difference between the two shapes becomes progressively smaller.

> ### 4. <u>Formulas for Degrees of Freedom vary by the Statistics and the test they are used in</u>

Statistic	df	Explanation
ANOVA: Mean Sum of Squares Within (MSW)	$N - k$	N: total # of all data points
ANOVA: Mean Sum of Squares Between (MSB)	$k - 1$	k: # of groups
χ^2	$n - 1$	n: Sample Size
χ^2 test for Goodness of Fit	$n - 1$	k: # of categories
χ^2 test for Independence	$(r - 1)(c - 1)$	# of rows and columns
χ^2 test for Variance	$n - 1$	n: Sample Size
F	$n_1 - 1$ and $n_2 - 1$	n_1 and n_2: Sizes of the 2 Samples
t	$n - 1$	n: Sample Size
1-Sample t-test, and Paired t-test	$n - 1$	
2 (Independent)-Sample t-test	$n_1 + n_2 - 2$	n_1 and n_2: Sizes of the 2 Samples

Related Articles in This Book: *Variance*; *Chi-Square – the Test Statistic and Its Distributions*; *Chi-Square Test for Goodness of Fit*; *Chi-Square Test for Independence*; *Chi-Square Test for the Variance*; *F*; *t – The Test Statistic and Its Distributions*; *t-tests – Parts 1 and 2*; *ANOVA—Part 3: One Way*

DESIGN OF EXPERIMENTS (DOE) – PART 1 (OF 3)

An exhaustive coverage of DOE is beyond the scope of this book. This 3-part series attempts to cover the key points and to clarify the most confusing concepts.

Summary of Keys to Understanding

 1. For a Process output y which is a function of several Factors (x's),
$$y = f(x_1, x_2, \ldots, x_n),$$
DOE can design the most efficient and effective experiments to determine the values of the x's which produce the optimal value of y.

 2. Since Designed Experiments provide strong evidence of Cause and Effect, DOE can also be used to validate – or invalidate – Regression Models.

 3. Statistical software packages perform DOE calculations which help to specify the elements which make up the Design: Levels, Combinations, Replications, Runs, Order.

 4. **Don't extrapolate.** Whatever **conclusions** we make as a result of the experiment **are only valid within the range of Levels tested**.

Our experiments show that 3 pills produce results which are 3 times as good as 1 pill.

So, 10 pills should be 10 times as good.

Actually, 10 pills would send you to the hospital.

 5. To start, identify all reasonably plausible Factors.

Explanation

> **1. For a Process output y, which is a function of several Factors (x's),**
>
> $$y = f(x_1, x_2, \ldots, x_n),$$
>
> **DOE can design the most efficient and effective experiments to determine the values of the x's which produce the optimal value of y.**

DOE can be used to select values for Factors (x Variables) which produce the optimal value for – or the minimal Variation in – the Response (y Variable). Examples of Factors in a laundry Process could be water temperature and type of detergent. The Response would be a measure of cleanliness.

DOE is active and controlling. (This can be done with Processes, but usually not with Populations).

DOE doesn't collect or measure existing data with pre-existing values for y and the x's. **DOE specifies Combinations of values for inputs (Factors) and then measures the resulting values of the outputs (Responses). This is the Design of the Experiment.**

DOE is more efficient and effective than other methods.

- Trial and Error is, by definition, chancy. It rarely gets good results.
- Testing one Factor at a time can require a large number of experimental Runs. **DOE uses statistics to minimize the number of Runs**. Also, 1-Factor-at-a-time does not **account for Interactions** between the Factors. **DOE does.** Interactions can be very important. We need to understand them.
- Testing all possible Factors at once can be inefficient and risky. The phased approach of **DOE allows for learning and adjusting during the experiment.**

> **2. Since Designed Experiments provide strong evidence of <u>Cause and Effect</u>, DOE can also be used to validate – or invalidate – Regression Models.**

Designed Experiments (those designed by DOE) provide much stronger evidence of Cause and Effect than Inferential Statistics. Designed Experiments are based on **careful statistical design and controlled conditions**.

Compared to Inferential Statistical analyses, **Designed Experiments are much less susceptible to unknown Factors outside the process.**

The article "Regression – Part 3" said that we can't use the same data to create a Regression Model and to test it. If a Regression Model is to be a valid model of Cause and Effect, it must be able to predict future data derived from controlled experiments. Experiments designed by DOE are the way to test this.

> **3. Statistical software packages perform DOE calculations which help to specify the elements which make up the Design:**
> - **Levels**
> - **Combinations**
> - **Replications**
> - **Runs**
> - **Order**

There are many calculations to be performed in the various aspects of DOE, and almost always, statistical software is employed to perform this work. So, we won't go into those calculations here.

Levels

Usually, for practical reasons (cost and time), **only two Levels of each Factor are tested**. Let's use a hypothetical example of a laundry experiment.

Levels can be high or low numerical values for Factors such as temperature or time. **Or they can represent Yes or No answers** to question such as "Was there a pre-wash cycle?", **or they can be a choice,** (e.g., "Detergent #1" or "Detergent #2").

We can use shorthand labels such as "−1" and "+1" (or just "−" and "+") to **identify the two Levels.** For example, "−1" for 50°F and "+1" for 150°F. (Later we'll see that this shorthand has a purpose, since we'll be "multiplying" the −1s and +1s.)

For numerical Levels, it is important to select values which are sufficiently separated to have measurably different effects. Often this means using values which are outside the normal operating range, sometimes considerably so.

Combinations

Let's say we have three Factors, each of which has two Levels (labeled "+1" and "−1").

Then there are $2^3 = 8$ possible **Combinations** of values which we can test: $(-1, -1, -1)$; $(-1, +1, -1)$; $(-1, -1, +1)$; $(-1, +1, +1)$; $(+1, -1, -1)$; $(+1, +1, -1)$; $(+1, -1, +1)$; and $(+1, +1, +1)$.

Replications

Every Process has some amount of random internal Variation ("noise"). If we repeat (Replicate) a given Combination 3 times, we are likely to get three somewhat different values of the Response. **DOE will specify a number of Replications which will enable this internal Variation to be accurately quantified** – and thus allow it to be separated from the Variation which is caused by varying the values of the Factors.

Runs

One "Run" is a single Combination tested once. Three Replications of one Combination is three Runs.

Order

Testing the Combinations in a random order is important, as we'll explain in Part 3. The software can provide the order in which the Combinations are to be tested.

 | **4. Don't extrapolate.** Whatever **conclusions** we make as a result of the experiment **are only valid within the range of Levels tested**.

Our experiments show that 3 pills produce results which are 3 times as good as 1 pill.

So, 10 pills should be 10 times as good.

Actually, 10 pills would send you to the hospital.

This is another reason to select low and high numerical Levels which are widely separated.

 | **5. To start, identify all reasonably plausible Factors.**

Statistics need not be involved in this. **Use subject matter expertise** (preferably from brainstorming with several knowledgeable people who have different roles in the process) to identify all reasonably plausible Factors which might influence the value of the Response, y. Typically, there would be 6 to 8 or more of these.

Please continue with the article, *Design of Experiments (DOE) – Part 2*.

DESIGN OF EXPERIMENTS (DOE) – PART 2 (OF 3)

Builds on the content of the article, Design of Experiments (DOE) – Part 1.

Summary of Keys to Understanding

 1. **The Estimated Effect of a Factor x_i is:**
E_i = (the Average of the y's in Runs where x_i was "High")
minus (the Average of the y's in Runs where x_i was "Low")

 2. **The "−1" and "+1" Coded Level notation is more than just a shorthand. These values can be multiplied to provide a formula for** Estimated Effect **of a Factor or Interaction.**

 3. **An Interaction is present when the two Levels of a Factor react differently to a change in Level of another Factor.**

Parallel lines indicate No Interaction. Crossed lines indicate an Interaction.

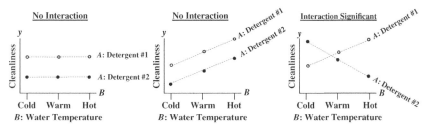

No Interaction	No Interaction	Interaction Significant

 4. **To calculate the Coded Level of an Interaction, AB for a given Run, multiply the Coded Level of A by the Coded Level of B.**

Run	A x_1	×	B x_2	=	AB $x_{1\times2}$
#1	−1		−1		+1
#2	+1		−1		−1
#3	−1		+1		−1
#4	+1		+1		+1
#5	−1		−1		+1
#6	+1		−1		−1
#7	−1		+1		−1
#8	+1		+1		+1

Explanation

> **1. The Estimated Effect of a Factor x_i is:**
> E_i = (the Average of the y's in Runs where x_i was "High")
> minus (the Average of the y's in Runs where x_i was "Low")

That seems like a common-sense way to do it.

Factors are the x Variables – the inputs – in the function, $y = f(x_1, x_2, \ldots, x_n)$. **The output, y, is called the "Response Variable."** (Sometimes, A, B, C, etc. are used instead of x_1, x_2, x_3, etc. to denote the Factors.)

In the simple example of 3 Factors (X_1, X_2, and X_3) and 2 Levels (low and high), we have $2^3 = 8$ different **Combinations** we can test. Without repeating a Combination, we can have 8 experimental Runs with 8 results for y.

Example

"-1" represents the low Level for the Factor, and "$+1$" represents the high Level. We call these "**Coded Levels.**"

Run	Inputs: Coded Levels of Factors				Output: Response value
	A x_1	B x_2	C x_3		y
#1	-1	-1	-1		1
#2	$+1$	-1	-1		6
#3	-1	$+1$	-1		6
#4	$+1$	$+1$	-1		11
#5	-1	-1	$+1$		2
#6	$+1$	-1	$+1$		11
#7	-1	$+1$	$+1$		5
#8	$+1$	$+1$	$+1$		13

The Estimated Effect of a Factor is the average value of y when the Factor is at a high Level minus the average value of y when the Factor was is at a low Level.

A was high in Runs 2, 4, 6, 8. Average y = (6+11+11+13)/4 = 10.25
A was low in Runs 1, 3, 5, 7. Average y = (1 + 3 + 4 + 5)/4 = 3.5
The Estimated Effect, E_1, of the Factor x_1, is 10.25 – 3.5 = **6.75**
Likewise, E_2, the Estimated Effect of x_2, is (35/4) – (20/4) = **3.75**
And E_3 is (31/4) – (24/4) = **1.75**

Terminology: The Estimated Effects of Factors – as opposed to Interactions between Factors – are called "**Main Effects.**"

2. The "−1" and "+1" Coded Level notation is more than just a shorthand. These values can be multiplied to provide a formula for the Estimated Effect of a Factor or Interaction.

For example, the Level for Factor x_1 was "High" in Runs 2, 4, 6, and 8. So, we multiply the y values for those Runs by the +1s that are in the x_1 column for those rows.

x_1 was Low in Runs 1, 3, 5, and 7. So, we multiply the y value by the −1s.

Level of x_1 is High

Run	x_1	×	y	=	
#2	+1		6		6
#4	+1		5		5
#6	+1		11		11
#8	+1		13		13
				Total:	41
		Divide by 4 for the **Average:**			**10.25**

Level is of x_1 is Low

Run	x_1	×	y	=	
#1	−1		1		−1
#3	−1		3		−3
#5	−1		4		−4
#7	−1		5		−5
				Total	−14
		Divide by 4 for the **Average:**			**−3.5**

(<u>Note</u>: the average value of the y's when x_1 is Low is 3.5. It is the multiplication by the −1s that turn it into −3.5)

The **Estimated Effect of** x_1, $E_1 = 10.25 + (−3.5) = 6.75$, just like we calculated above. If we do this in a single table, we can get the same result by single addition of the values of the rightmost column and then divide by half the number of Runs (4).

Run	x_1	×	y	=	
#1	-1		1		-1
#2	$+1$		6		6
#3	-1		6		-6
#4	$+1$		11		11
#5	-1		2		-2
#6	$+1$		11		11
#7	-1		5		-5
#8	$+1$		13		13
				Total:	
Divide by 4 for E_1, the **Effect of** x_1:					**6.75**

This series of calculations can be summarized in a formula:

$$E_i = \frac{1}{2^{k-1}} \sum_{j=1}^{2^k} c_{i,j} y_j \quad \text{or} \quad E_i = \frac{\sum_{j=1}^{2^k} c_{i,j} y_j}{2^{k-1}}$$

where

E_i is the estimated Effect of the Factor x_i

i is the number identifier of the Factor

k is the number of Factors *(3, in this example)*;

2^k is the number of Runs;

2^{k-1} is half the number of Runs

j is number of the Run *(from 1 to 8 in this example)*

c_{ij} is the Coded Level ("-1" or "$+1$") for "Low" or "High" Levels of the Factor i in Run j; y_j is the value of the Response y in the Run j.

> **3. An <u>Interaction</u> is present when the two Levels of one Factor <u>react differently</u> to a change in Level of another Factor.**

For example, we are trying to maximize cleanliness in a laundry Process. y is a numerical measure of cleanliness

Factor 1, A, is detergent type. The Levels are detergent #1 and detergent #2

Factor 2, B, is brand of washing machine. The Levels are Brand P and Brand Q

Factor 3, C, is the amount of bleach added. The levels are Low and High

Factor 4, D, is water temperature. The levels are Low and High

Left Diagram below:

– Detergent #1 cleans better than Detergent #2.

– However the brand of washing machine makes no difference.

So, the two Levels of detergent type react the same (no reaction at all) to a change in the Level of washing machine. Thus, there is no Interaction between x_1 and x_2.

Center Diagram:

– An increase in the Level of Factor C, bleach, increases the value of Y by an equal amount for the two Levels of detergent. There is no synergy between bleach and detergent type. There is no Interaction.

Parallel lines indicate No Interaction. Crossed lines indicate an Interaction.

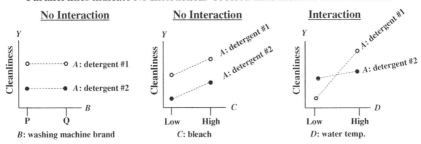

Right Diagram:

One Level (Detergent #1) of Factor A (detergent type) reacts significantly stronger than the other Level (Detergent #2) to a Level change in Factor D (water temperature). There is an Interaction. When the lines are crossed at substantially differently angles – as in the right diagram – the Effect of the Interaction is likely to be Statistically Significant. If the lines cross, but are close to parallel, the Interaction may not be Statistically Significant. Statistical software will tell us which is the case.

 4. **To calculate the Coded Level of an Interaction, AB, simply multiply the Coded Level of A by the Coded Level of B.**

Run	A x_1	×	B x_2	=	AB $x_{1×2}$
#1	−1		−1		+1
#2	+1		−1		−1
#3	−1		+1		−1
#4	+1		+1		+1
#5	−1		−1		+1
#6	+1		−1		−1
#7	−1		+1		−1
#8	+1		+1		+1

Performing these multiplications for all Interactions for 3 Factors, we get the following table:

Run	A x_1	B x_2	C x_3		AB $x_1 x_2$	AC $x_1 x_3$	BC $x_2 x_3$		ABC $x_1 x_2 x_3$
	c_1	c_2	c_3		c_{12}	c_{13}	c_{23}		c_{123}
#1	−1	−1	−1		+1	+1	+1		−1
#2	+1	−1	−1		−1	−1	+1		+1
#3	−1	+1	−1		−1	+1	−1		+1
#4	+1	+1	−1		+1	−1	−1		−1
#5	−1	−1	+1		+1	−1	−1		+1
#6	+1	−1	+1		−1	+1	−1		−1
#7	−1	+1	+1		−1	−1	+1		−1
#8	+1	+1	+1		+1	+1	+1		+1

As before, the top row of these tables shows two different naming conventions for each Factor and Interaction. For example, "*AB*" and "$x_1 x_2$" are two different names for the same Interaction. In this table, we added another row to show how the Coded Levels (the c's) are denoted. For example, the name of the Coded Levels for *AB* (for any Run) will include "c_{12}." The name for the Coded Level of *AB* in the *j*th Run will be of the form "$c_{12,j}$." For example, $c_{12,8} = +1$

This table includes three different sets of information:

Design: The Runs and the Factor columns, *A*, *B*, *C*, tell us how to set the inputs (Levels of Factors) for each Run. We know this prior to running the experiment. In this example, there is just one Replication of the Runs. Each Run is to be performed in the order shown. The order of the Runs is set by statistical software to ensure Randomness.

Calculated Levels: We cannot set Levels for Interactions; **the Interaction Columns are calculated from the Factor Columns**. We could calculate them prior to the experiment, but that information is not useful without the *y* values which are measured during the experiment.

Notation:

- It seems that, for labelling Interactions, the letter notation (ABC) is more common – and maybe less confusing – than using *x*'s with subscripts.

- On the other hand subscripts are useful in formulas that involve summation signs.

For example, in the Formula for Estimated Effects, the coded value for the *ABC* Interaction in the 7th run is $c_{123,7}$

– Here, we are using c's with subscripts to denote the Coded Levels ("−1" or "+1"). Other publications sometimes use c's with subscripts to denote the estimate of Effects (instead of E's with subscripts, as we do here.)

We can use these Coded Levels and our formula to calculate the Effect for each Factor and each Interaction. For example, Factor *AB* (aka $x_1 x_2$) has the Coded Levels $\mathbf{C_{12,j}}$ where *j* goes from 1 to 8.

$$E_{12} = \frac{1}{2^{3-1}} \sum_{j=1}^{8} c_{12,j} y_j$$

$$= \frac{1}{4}[(+1)(1) + (-1)(6) + (-1)(6) + (+1)(11) + (+1)(2)$$
$$+ (-1)(11) + (-1)(5) + (+1)(13)]$$

$$= \frac{1}{4}[-1] = -0.25$$

Other than zero, this is the smallest Effect a Factor can have, so it is unlikely that it is Statistically Significant. The statistical software will tell us for sure.

=> Please continue with the article "Design of Experiments (DOE) – Part 3"

DESIGN OF EXPERIMENTS (DOE) – PART 3 (OF 3)

Builds on the content of the articles, Design of Experiments – Parts 1 and 2.

Summary of Keys to Understanding

1. <u>Full Factorial designs</u> test all possible Combinations of Factors and their Levels. They yield the best information on the Effects of all Factors and all possible Interactions. But, they can be prohibitively expensive.

2. <u>Fractional Factorial designs</u> test fewer Combinations and can provide acceptable rigor.

Full Factorial for 2 Factors			⇨	Calculate Interaction for Confounding			⇨	Fractional Factorial for 3 Factors		
Run	A x_1	B x_2		**Run**	A x_1	B x_2	AB x_1x_2		**Run**	A x_1	B x_2	$C{\approx}AB$ x_3
#1	−1	−1		#1	−1	−1	+1		#1	−1	−1	+1
#2	+1	−1		#2	+1	−1	−1		#2	+1	−1	−1
#3	−1	+1		#3	−1	+1	−1		#3	−1	+1	−1
#4	+1	+1		#4	+1	+1	+1		#4	+1	+1	+1

3. <u>DOE Screening Experiments</u> are Resolution III or IV experiments **designed to tell us which Factors are most Significant.**

4. Next, Full Factorial or Fractional <u>Factorial Experiments</u> are designed and conducted, using just the most Significant Factors which were identified in the Screening Experiments. The focus is on <u>Interactions</u>, as well as Main Effects.

5. For Nuisance Factors and Unknown Factors,
      Block (group) what you can; Randomize what you can't.

6. Analyze the data and confirm the DOE results.

Explanation

> **1. Full Factorial designs test all possible Combinations of Factors and their Levels. They yield the best information on the Effects of all Factors and all possible Interactions. But, they can be prohibitively expensive.**

In most DOE experiments, only two Levels — "High" and "Low" – are tested. If k is the number of Factors, then there are 2^k different Combinations. A 2^k design is called a **Full Factorial design.**

In Part 1, the first step was to identify 6 to 8 or more possible Factors to be evaluated. Six factors would yield $2^6 = 64$ Combinations to be tested – each with several Replications. **That many Runs would give us the needed information to analyze all Factors and all possible Interactions**.

There would be 2-level Interactions, such as AB, 3-level Interactions, like ABC – all the way up to the 6-level Interaction, $ABCDEF$.

However, performing that many Runs would often be prohibitively expensive or otherwise impractical.

> **2. Fractional Factorial designs test fewer Combinations and can provide acceptable rigor.**

If a Full Factorial design has 2^k Combinations, then a Fractional Factorial design has a fraction of that number 2^{k-p} Combinations. If we set $p = 1$, then we'll get a half-fractional design, because 2^{k-1} is half of 2^k. For our simple example, let's have $k = 3$ (indicating 3 Factors), and $p = 1$. So, we'll have $2^2 = 4$ Combinations, instead of $2^3 = 8$.

We start with a Full Factorial table (left) of Coded Levels for $k - 1 = 2$ of our 3 Factors. We can call the Factors A and B or x_1 and x_2. With 2 Factors, we have 4 Combinations and 4 Runs.

Full Factorial for 2 Factors			⇨	Calculate Interaction for Confounding				⇨	Fractional Factorial for 3 Factors			
Run	A x_1	B x_2		Run	A x_1	B x_2	AB x_1x_2		Run	A x_1	B x_2	$C≈AB,$ x_3
#1	−1	−1		#1	−1	−1	+1		#1	−1	−1	+1
#2	+1	−1		#2	+1	−1	−1		#2	+1	−1	−1
#3	−1	+1		#3	−1	+1	−1		#3	−1	+1	−1
#4	+1	+1		#4	+1	+1	+1		#4	+1	+1	+1

Next (center table), we'll multiply the two columns of these Coded Levels to calculate a third column, which will therefore have Coded Levels for the Interaction *AB*.

Step 3 is to re-label the Interaction column as the column for the **third Factor, *C*, mixed up or Confounded with the Interaction of *A* and *B*. So we're using the design calculated for the Interaction column in testing for the third Factor.** This Fractional Factorial design has 4 Runs for 3 Factors instead the 8 Runs which the Full Factorial design would have.

This would be acceptable so long as we have reason to believe that the Effect of the Interaction is not Statistically Significant.

- **Not infrequently, 2-level Interactions have Significant effects**, and these can be very useful in explaining apparently strange process outcomes.
- **Significant 3-level Interactions are comparatively rare.**
- **There is some difference in opinion regarding Significant 4-Level Interactions**, that is, are they so exceedingly rare as to be ignored?
- **5-or more–Level Interactions are not worth considering**.

Our simple example of a 3-Factor Fractional Factorial (2^{3-1}) design would be risky. We're Confounding a Main Effect (*C*) with a 2-Factor Interaction (*AB*), which could be Significant.

This $C \approx AB$ Confounding is actually the design **Generator** for an **Alias Structure** which the statistical software will produce. This structure may involve Confounding other Factors with Interactions and Interactions with each other. *(Terminology note: "Aliasing and Aliased" are sometimes used as synonyms for "Confounding" and "Confounded.")*

Reviewing this Alias structure may reveal, for example, that a specific Factor, say *B*, is involved with all of the Confounded 2-Factor Interactions. **We, not the statistical software, choose which real-world Factors we assign to which letters (*A, B, C*, etc.).** So, we should designate as "*B*" a real-life Factor which is unlikely to Interact with others.

In the laundry process described in the Part 2 article, we might expect that the Factor washing machine brand would not interact with other Factors, so we could designate that as Factor *B*. Then, the aliasing structure could Confound *B* with other the Factors involved in 2-Factor Interactions, and we could greatly lessen our concern that these Interactions would be Statistically Significant.

 3. DOE Screening Experiments are Resolution III or IV experiments **designed to tell us which Factors are most Significant.**

In Part 1, we said the first step was to use subject matter expertise to identify all possible or reasonably plausible Factors – 6 to 8 or more of these.

We would like to save our time and budget focusing on the "critical few" Factors with the most impact. So, Step 2 is to use a "Screening Experiment" to screen out Factors that are not Statistically Significant or which have the least significance. So, we don't need as high a "Resolution" experiment as we will later. Resolution III or IV is sufficient.

Resolution measures the extent to which estimated Main Effects are Confounded with Interactions.

Resolution III: (1+2)

Main Effect (1) Confounded with 2-Factor Interactions

Resolution IV: (1+3 or 2+2)

Main Effect (1) Confounded with 3-Factor interactions and

2-Factor interactions Confounded with other 2-Factor Interactions

Statistical software can serve up a choice of design options of various Resolutions for us

Terminology note: "2-Factor" and "3-Factor," etc. Interactions are sometimes called "2nd Order" and "3rd Order" etc. Interactions.

In the ANOVA table output from the Screening Experiment,

ANOVA table						
Factors	df	SS	MS	F	Effect	p
A Detergent	1	91.1	91.1	729	6.75	0.02
B Water Temp.	1	28.1	28.1	225	3.75	0.04
C Washing Machine	1	6.1	6.1	49	1.75	0.09
Interactions						
AB Detergent × Water Temp.	1	0.1	0.1	1	−0.25	0.50
AC Detergent × Washing Machine	1	6.1	6.1	49	1.75	0.09
BC Water Temp × Washing Machine	1	3.1	3.1	25	−1.25	0.13
ABC Detergent × Temp X Machine	1	0.1	0.1	1.0	−0.3	0.5

- **a Factor with a $p \leq \alpha$ (α is most commonly selected to be 0.05) has a Statistically Significant Effect on the Response (y).** We'll screen out those that don't. In the table above, A (detergent) and B: (water temperature) have Statistically Significant Effects. C (washing machine brand) does not. So, we would keep A and B, and proceed without C.

- ANOVA is used to apportion the Variation contributed by each Factor and Interaction. The **Sums of Squares (SS) column indicates the relative magnitude of the Effect of each Factor** on the Response. We may want to proceed with just the top 2 or 3 Significant Factors – if there is sizeable gap in Sums of Squares between these and the others. In this example, we have only two Significant Factors (*A* and *B*), so we don't need to use the SS for screening purposes. A Pareto chart can also be used to determine which Factors survive the Screening Experiment and move on to the next step.

If none of the Factors have $p \leq \alpha$:

- The original list may not have included Factors that had a Significant influence, and/or
- The Levels were not sufficiently separated to result in a difference in y.

> **4. Next, Full Factorial or Fractional <u>Factorial Experiments</u> are designed and conducted, using just the most Significant Factors which were identified in the Screening Experiments. The focus is on <u>Interactions</u>, as well as Main Effects.**

In a Screening Experiment, we ignore Interactions, but **Interactions can have very important effects on Responses.** So our next set of experiments will focus on both Interactions and Main Effects.

Earlier, we said that there were $2^3 = 8$ possible Combinations for 2 Levels of 3 Factors. A Full Factorial experiment tests all possible Combinations. For 2 Levels of k factors, there are 2^k possible Combinations. If we tested all of them, we would have a Full Factorial Design.

Screening Experiments are often Fractional Factorial Experiments which fully test fewer than 2^k Combinations. **If time and budget permit, at this stage, we would perform a Full Factorial Experiment** with the Factors selected in the Screening Experiment. If we use our Screening Experiment to reduce the number of Significant Factors to 3 or 2, then we are more likely to be able to afford the Full Factorial number of Runs.

With 4, 5 or more Factors at this point, budget and time constraints could require a Fractional Factorial design. **The statistics software can calculate the Resolution for various options. We select the option with which we are comfortable,** based on the following definitions.

Resolution IV: (1+3 or 2+2)

- **Main Effects Confounded with 3-Factor Interactions**. (Significant 3-Factor Interactions are fairly rare, so this may be OK for some experiments.)
- 2-Factor interactions Confounded with other 2-Factor Interactions

Resolution V: (1+4 or 2+3 or 3 + 2)

- **Main Effects Confounded with 4-Factor Interactions**. (Significant 4-Factor Interactions are almost unheard of, so **this Resolution should be fine for almost all purposes.**)
- 2-Factor Interactions Confounded with 3-Factor Interactions
- 3-Factor Interactions Confounded with 2-Factor Interactions

Resolution VI: Main Effect Confounded with 5-Factor Interactions. This is overkill. Stick with Resolution V.

Or, since this is only possible with 5 or more Factors, use your Screening Experiment to select the top 4 Significant Factors.

> **5.** For Nuisance Factors and Unknown Factors,
> **Block (group) what you can; Randomize what you can't.**

A Nuisance Factor is one outside the Process. This is also known as a Special Cause of Variation. For example, the ambient temperature of a factory can increase steadily as the day goes on. For some Processes, this can affect the results. (y values). See the article *Control Charts – Part 1* for more on Special Cause Variation.

A known Nuisance Factor can often be Blocked. **To "Block" in this context means to <u>group</u> into Blocks.** By so doing, we try to remove the Effect of Variation of the Nuisance Factor. In this example, we Block the Effect of the daily rise in ambient temperature by performing all our experimental Runs within a narrow Block of time. And, if it takes several days to complete all the Runs, we do them all at the same time of day and the same ambient temperature. **We thus minimize the Variation in the Nuisance Factor**, ambient temperature. **That minimizes the Variation in y caused by the Nuisance Factor.**

There can also be Factors affecting y which we don't know about. Obviously, we can't Block what we don't know. But we can often avoid the influence of Unknown Factors (also known as "Lurking" Variables) by Randomizing the order in which the experimental combinations are tested.

For example – unbeknownst to us – the worker performing the steps in an experiment may get tired over time, or, conversely, they might "get in a groove" and perform better over time. **So we need to Randomize the order in which we test the Combinations. Statistical software can provide us with the random sequences to use in the experiment.**

 | **6. Analyze the data and confirm the DOE results.**

We said in the Part 1 article that a full coverage of DOE is beyond the scope of this book. That certainly holds for the analysis of the results of the experiments. Here are some of the things that should be done. Use subject matter knowledge in addition to the statistical tools.

- Find the Statistically Significant Factors and Interactions ($p \leq \alpha$).
- Re-run the analysis with only these Statistically Significant Effects.
- Analyze all the data and graphs produced by the software; look for anomalies, time-order effects, etc.
- Create a Regression Model from the data; R^2 should be high.
- Residual plots from the Regression Model should be Normally distributed around zero.
- Find the optimal settings of the Factors.
- Run several tests at the optimal settings to confirm the results. The Regression Model can only be proven valid if it correctly predicts future results.

Related Articles in This Book: *Design of Experiments (DOE) – Parts 1 and 2*; *ANOVA, Parts 1 – 4*; *Sums of Squares*; *Alpha (α)*; *p-Value, p*; *Control Charts – Part 1: General Concepts and Principles*; *Regression Parts 1–5*; *Residuals; Charts, Graphs, Plots – Which to Use When*

DISTRIBUTIONS – PART 1 (OF 3): WHAT THEY ARE

 1. A Distribution (also known as Probability Distribution) is a set of **values of a Variable, along with the associated Probability of each value of the Variable.** Distributions are usually plotted with the Variable on the horizontal axis and the Probability on the vertical axis.

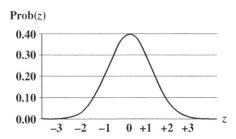

 2. Named Distributions usually occur in families, e.g., Normal Distributions, t-Distributions, F-Distributions, Binomial Distributions, etc.

 3. Different Distributions can have Discrete or Continuous Probability Curves for Discrete or Continuous data.

Distribution	Data	Probability Curve
Binomial, Hypergeometric, Poisson	Discrete	Discrete
Exponential, Normal, t	Continuous	Continuous
F, Chi-Square	Both	Continuous

 4. Distributions can be numerically described by three categories of Parameters: Central Tendency (e.g., Mean), **Variation/Spread** (e.g., Standard Deviation), **Shape** (e.g., Skew).

Central Tendency	Variation/ Spread	Shape

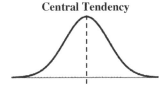

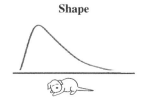

Explanation

1. **A Distribution** (also known as Probability Distribution) is a set of **values of a Variable, along with the associated Probability of each value of the Variable.** Distributions are usually plotted with the Variable on the horizontal axis and the Probability on the vertical axis.

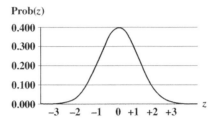

This is a graph of the Standard Normal Distribution, whose Variable is z. We can see that the Probability of $z = 0$ is about 0.4 and the Probability of 1 or -1 is about 0.2.

2. **Named Distributions usually occur in families**, e.g., Normal Distributions, t- Distributions, F-Distributions, Binomial Distributions, etc.

Some authors say that a Distribution is any collection of data values for a Variable. But that could just as easily describe a Sample or a Population, and it is not descriptive of how Distributions are generally used in statistics. So, here we will focus on <u>named families</u> of Distributions.

The Standard Normal Distribution in the graph above is a member of the Normal Distribution family. Different values for the Mean and/or Standard Deviation would produce different members of the family with different Probabilities. For the Normal Distribution, the shapes would still be fairly similar. But different values for the defining or generating properties of other families can result in dramatic differences as shown below:

<u>3 members of the Binomial Distribution family</u>

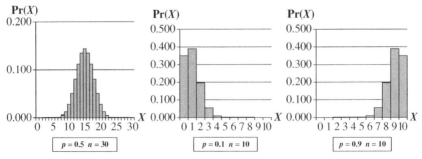

Distribution families and the properties which generate different Distributions within each family:

- Binomial: number of trials (n) and Probability of each trial (p)
- Chi-Square: Degrees of Freedom (df)
- Exponential: Mean (μ)
- F: Degrees of Freedom (df) for the numerator and df for the denominator
- Hypergeometric; Proportion of successes
- Normal: Mean (μ) and Standard Deviation (σ)
- Poisson: Mean (μ, also known as λ, the Expected Value)
- t: Degrees of Freedom (df)

There are a number of named families of Probability Distributions. This 3-part series of articles on Distributions will concentrate on these eight commonly-used ones. There are individual articles on each. Four of these Distributions have their own Test Statistics: Chi-Square, F, t, and Normal (z).

3. Different Distributions can have Discrete or Continuous Probability Curves for Discrete or Continuous data.

Distribution	Data	Probability Curve
Binomial, Hypergeometric, Poisson	Discrete	Discrete
Exponential, Normal, t	Continuous	Continuous
F, Chi-Square	Both	Continuous

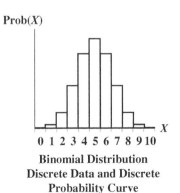

Prob(X)

Binomial Distribution
Discrete Data and Discrete
Probability Curve

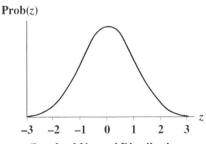

Prob(z)

Standard Normal Distribution
Continuous Data and Continuous
Probability Curve

These graphs show the difference between a Distribution that has a Discrete data and a Discrete curve compared to a Distribution with Continuous data and a Continuous curve.

For the Discrete data Distribution, the values of the Variable X can only be non-negative integers, because they are Counts. There is no Probability shown for 1.5, because 1.5 is not an integer, and so it is not a legitimate value for X. **The Probabilities for Discrete data Distribution are shown as separate columns.** There is nothing between the columns, because there are no values on the horizontal axis between the individual integers.

For Continuous Distributions, values of horizontal-axis Variable are real numbers, and there are an infinite number of them between any two integers. Continuous data are also called Measurement data; examples are length, weight, pressure, etc. **The Probabilities for Continuous Distributions are infinitesimal points on smooth curves.**

For the first six Distributions described in the table above, the data used to create the values on the horizontal axis come from a single Sample or Population or Process. And the data are either Discrete or Continuous. **The F and Chi-Square (χ^2) Distributions are hybrids.** Their horizontal axis Variable is calculated from a ratio of two numbers, and the source data don't have to be one type or another. (This is explained in the three articles on the different types of Chi-Square tests.) Being a ratio, the horizontal axis Variable (F or χ^2) is Continuous. The Probability curve is smooth and Continuous.

 | **4. Distributions can be numerically described by three categories of Parameters: Central Tendency** (e.g., Mean), **Variation/Spread** (e.g., Standard Deviation), **Shape** (e.g., Skew).

The named Distributions are intended to represent Populations (and not usually Samples), so we use the term "Parameter" (instead of "Statistic") to describe measures of their properties. There are three categories of Parameters which we can use to describe a Distribution:

- **Central Tendency**: e.g., Mean, Mode, Median
- **Variation** (aka "Variability," "Dispersion," "Spread," and "Scatter") e.g., Standard Deviation, Variance, Range
- **Shape:** e.g., Skew and Kurtosis

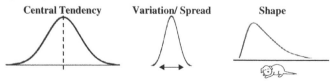

Central Tendency   Variation/ Spread   Shape

Central Tendency

The Mean is the average of all the x values. The Median is the x value in the middle of the range, and the Mode is the x value that is most common.

In the Poisson Distribution below left, the Mean is four, and the Modes are three and four. Note that – unlike Mean and Median – there can be more than one Mode. And the two or more Modes do not need to be contiguous, as they are in this example. In the Exponential Distribution below right, the Mean is five, and the Mode is zero. In both cases, the Median is not meaningful, since the Range extends indefinitely toward the right.

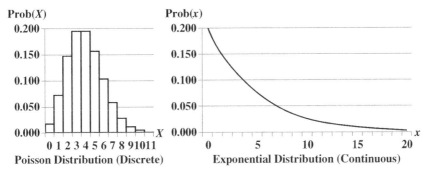

Poisson Distribution (Discrete)   Exponential Distribution (Continuous)

Actually, for these two Distributions, we didn't need to calculate the Mean; the Mean is the Parameter which is used to generate individual Distributions within these two families. **The individual members of the Poisson and Exponential Distribution families are produced by their individual values for the Mean.** Similarly, **individual members of the Normal family are defined by their individual values for two Parameters – Mean and Standard Deviation**. For example, the Standard Normal Distribution is the one whose Mean is zero and Standard Deviation is 1.

In statistical analyses like t-tests, ANOVA, and ANOM, the Means of Samples or Populations (or a Sample and a Population) are compared as a way of determining whether the two entities are statistically similar or different. In Nonparametric analyses, where data do not approximate a Normal Distribution, the Medians are compared.

Variation/Variability/Dispersion/Spread/Scatter

All mean the same thing, and there is a separate article devoted to that subject. They are measures of how "spread-out" a Distribution is. The most useful measure is Standard Deviation. Other measures include Variance, Range, and InterQuartile Range. These are all explained in that article.

Shape

Skew or "Skewness" and Kurtosis are the two most commonly used measures of Shape. If a Distribution is left–right symmetrical, it has a Skew of zero. Otherwise, the **Skewness measures the direction and the degree to which the Distribution appears to be stretched out in one direction or another.** See the article *Skew, Skewness*.

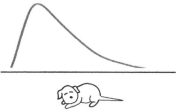

Tail is a term used to describe the rightmost and/or leftmost parts of a Distribution. The dog's tail in the picture above points in the direction of the Skew. Since the right tail is thicker across a longer stretch in the illustration above, the Skewness is positive.

Kurtosis is a measure of how "pointy" the Distribution is. The Normal Distribution has a Kurtosis of zero. Distributions pointier than the Normal have a positive Kurtosis. Less-pointy Distributions (which then have heavier tails) have a negative Kurtosis.

Related Articles in This Book: *Distributions – Part 2: How They Are Used*; *Distributions – Part 3: Which to Use When*; *Which Statistical Tools to Use to Solve Some Common Problems*; *Binomial Distribution*; *Chi-Square – the Test Statistic and Its Distributions*; *Exponential Distribution*; *F – the Test Statistic and Its Distributions*; *Hypergeometric Distribution*; *Nonparametric*; *Normal Distribution*; *Poisson Distribution*; *Skew, Skewness*; *t, the Test Statistic and its Distributions*; *Test Statistic*; *Variation/Variability/Dispersion/Spread*

DISTRIBUTIONS – PART 2 (OF 3): HOW THEY ARE USED

 1. A Discrete Distribution – like the Binomial, Hypergeometric, or Poisson **– can provide Probabilities for individual values of X or Cumulative Probabilities of a range of values.**

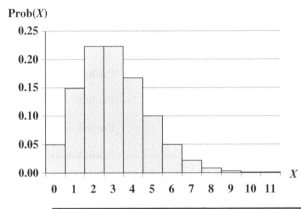

Prob(X)

Prob($X = 0$): **0.050**

Prob($X = 1$): **0.149**

Prob($X = 2$): **0.224**

Prob($X = 3$): **0.224**

Prob($X = 4$): **0.168**

Prob($X = 5$): **0.101**

 2. A Continuous Distribution associated with a Test Statistic – e.g., the Standard Normal, t, F, and Chi-Square – **can take a value for the Cumulative Probability, Alpha, and give us a Critical Value** of the Test Statistic.

 3. Or it can take a calculated value of the Test Statistic and give us a Cumulative Probability, the p-value.

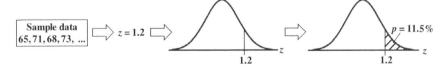

 4. Then, we can compare p to α or, (equivalently) the Critical Value to the value of the Test Statistic to determine the outcome of a Hypothesis Test. Alternately, we can use the Critical Value in the Confidence Interval method.

 5. Which Distribution to use is dependent on the type of data and the type of test or analysis. That is the subject of the Part 3 article.

171

Explanation

> **1. A Discrete Distribution** – like the Binomial, Hypergeometric, or Poisson **– can provide Probabilities for individual values of X or Cumulative Probabilities of a range of values.**

In all Distributions, the vertical axis is the Probability of a value on the horizontal axis occurring. In Discrete Distributions – like the Binomial, Hypergeometric, or Poisson – the horizontal axis is Discrete data representing Counts.

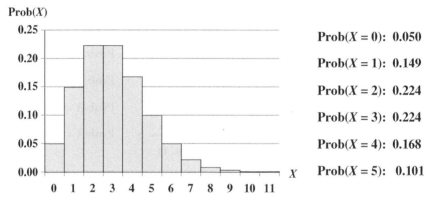

Rather than the individual Probability of a specific value, say X, of the horizontal axis Variable, **we are usually more interested in the Probability of all values greater than or less than X.** This kind of Probability of a range of values is called a Cumulative Probability.

For a Discrete Distribution, a Cumulative Probability is just the sum of the individual Probabilities of the Counts within a range.

For example, let's say we are operating a small call center which averages 3 incoming calls every 10 minutes. We can handle up to 5 calls in 10 minutes. What is the Probability that our capacity will be exceeded in any given 10 minute period?

The Poisson Distribution above can tell us the Probability of exactly $X = 6$ in 10 minutes. But, we also need to know the Probabilities of 7, 8, and so on. The Poisson Distribution can tell us those also.

But, since there is theoretically no limit on the number of calls, our approach is to use the Poisson Distribution to get the Probabilities of $X = 0, 1, 2, 3, 4$, and 5. We total these and subtract from 1 (or 100%) to get the Probability of exceeding 5.

The Probability of X being five or fewer is the sum of the six Probabilities shown to the right of the diagram: 0.916. So the Probability of exceeding our limit of 5 calls is $1 - 0.916 = 0.084$.

For Continuous Distributions, it is a little more complicated, since there are an infinite number of point Probabilities in any range of horizontal axis values. If we **consider a Continuous Distribution to be the limiting case of a Discrete Distribution with narrower and narrower columns**, we know from calculus that **the integral would play the role of summing the values of the (infinitely narrow) columns**. We also know that **the integral is the area under the curve** above the specified range of values on the horizontal axis.

Cumulative Probabilities

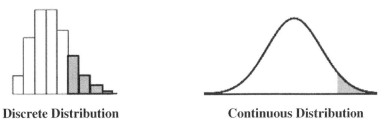

Discrete Distribution   **Continuous Distribution**

 | **2. A Continuous Distribution associated with a Test Statistic** – e.g., the Standard Normal, t, F, and Chi-Square – **can take a value for the Cumulative Probability, Alpha, and give us a Critical Value** of the Test Statistic.

In a Continuous Distribution, the horizontal axis Variable is not a Count; it is a measurement, like a test score. Continuous Distributions are used differently. They are used in Inferential Statistics, in which we take a Sample of data, and then we calculate a Statistic, like the Sample Mean. We then use this Statistic as an estimate of the corresponding property of the Population or Process from which the Sample was taken.

Inferential Statistics involve the concepts of Alpha, p, Critical Value, and Test Statistic. And Distributions play an integral part in determining the values of these things.

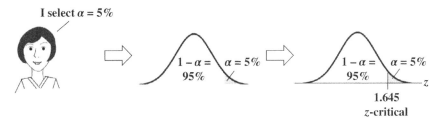

Before collecting a Sample of data, we must select a value for Alpha, the Level of Significance. If we want a 95% Confidence Level of not getting an Alpha Error, we select $\alpha = 5\%$. We then plot that as a shaded area under the curve of the appropriate Continuous Distribution. The boundary point separating the shaded from the unshaded area is the Critical Value. In this case we are using the Standard Normal Distribution, which has z as its Test Statistic. Tables or software can tell us that 5% (for a right-tailed test) corresponds to a Critical Value for of 1.645.

> **3. Or it can take a calculated value of the Test Statistic and give us a Cumulative Probability, the p-value.**

We just showed how we can start with a Cumulative Probability and use a Distribution to get a point value which serves as its boundary. We can do the opposite also: start with a point value and use the Distribution to give use the area under the curve (Cumulative Probability) beyond that point.

We take our Sample of data, and – using a formula for the Test Statistic – we calculate a point value for the Test Statistic, $z = 1.2$ in this case. Tables or software for the Distribution tell us that 11.5% is the corresponding Cumulative Probability.

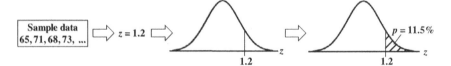

> **4. Then, we can compare p to α or, (equivalently) the Critical Value to the value of the Test Statistic to determine the outcome of a Hypothesis Test. Alternately, we can use the Critical Value in the Confidence Interval method.**

Since p is derived from z, and since Alpha determines the value of z-critical, comparing p to Alpha is statistically identical to comparing z to z-critical. So, we can use either comparison in coming to our conclusion.

Hypothesis Testing and Confidence Intervals are the two main methods of Inferential Statistics.

Hypothesis Testing

The diagrams in the table below are close-ups of the right tail of the Distribution in a right-tailed test.

Areas under the curve (right tail) $\alpha:\boxed{}$ $p:\boxed{\!\!\diagonal\!\!}$	z-critical $\overset{\text{\tiny l}}{\underset{z}{}}$	$\overset{\text{\tiny l}}{\underset{z}{}}$ z-critical
	$p \le \alpha$ $z \ge$ z-critical	$p > \alpha$ $z <$ z-critical
The observation from the Sample data is an accurate estimate for the Population or Process as a whole.	True	False
Null Hypothesis	Reject	Accept (Fail to Reject)
The observed difference, change, or effect is:	Statistically Significant	not Statistically Significant

Confidence Intervals

The Distribution and Alpha are used to determine the bounds of a Confidence Interval, as illustrated below for a 2-tailed test.

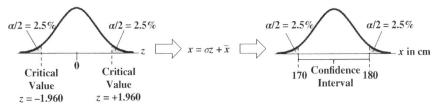

The articles on Confidence Intervals explain how they are used in coming to a conclusion about the test.

   5. Which Distribution to use is dependent on the type of data and the type of test or analysis to be used.

This is explained in the following article, Distributions – Part 3: Which to Use When

Related Articles in This Book: *Distributions – Part 1: What They Are; Distributions – Part 3: Which to Use When; Alpha, p-Value, Critical Value, and Test Statistic – How They Work Together; p, t, and F: ">" or "<" ?; Critical Values; Confidence Intervals – Parts 1 and 2; Hypothesis Testing – Parts 1 and 2*

Plus, there are these articles on individual Distributions:

- *Binomial Distribution*
- *Chi-Square – the Test Statistic and its Distributions*
- *Exponential Distribution*
- *F*
- *Hypergeometric Distribution*
- *Normal Distribution*
- *Poisson Distribution*
- *t – the Test Statistic and Its Distributions*
- *z*

DISTRIBUTIONS – PART 3 (OF 3): WHICH TO USE WHEN

There are individual articles in this book devoted to each of these Distributions.

			Distribution
Continuous Data, Continuous Distribution			
Compare 2 Means			
	Population or Process Standard Deviation is not known		*t*
	Population or Process Standard Deviation is known.		
		Sample Size < 30	*t*
		Sample Size ≥ 30	*t* or **Normal** (*z*)
Compare Variances			
	Two Sample Variances		*F*
	Sample Variance to specified Variance		**Chi-Square**
Involves time to an event or between events			**Exponential**
Discrete/Count Data, Discrete Distribution "What is the Probability of … ?"			
Occurrences are counted			**Poisson**
Units are counted			
	Sampling Without Replacement		**Hypergeometric**
	Sampling With Replacement, and other criteria met		**Binomial**
Discrete/Count Data, Continuous Distribution			
Compare Observed to Expected Counts			**Chi-Square**
Compare 2 or more Proportions			**Chi-Square**
Compare 2 Proportions			**Normal** (*z*)

See the Part 1 article for an explanation of Continuous vs. Discrete data and Distributions.

See also the article, "Which Statistical Tool to Use to Solve Some Common Problems."

ERRORS – TYPES, USES, AND INTERRELATIONSHIPS

Summary of Keys to Understanding

 1. Errors in statistics can be classified into two kinds:
- **Experimental Errors:** someone or something did something wrong.
- **Sampling Errors:** a statistical estimate from a Sample is not identical to the property of the Population or Process which it estimates.

 2. Experimental Errors largely involve errors in collecting the data – Measurement Errors and Sampling Bias.

 3. Sampling Errors include Alpha and Beta Errors, Margin of Error, Regression Residuals, and Sum of Squares Error.

 4. Some types of Sampling Errors influence each other.

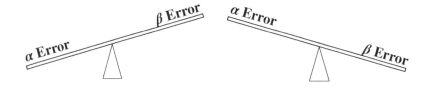

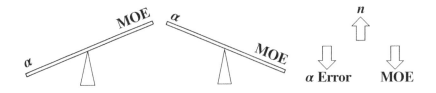 **5. An increase in Sample Size reduces Sampling Errors.**

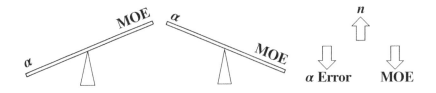

Statistics from A to Z: Confusing Concepts Clarified, First Edition. Andrew A. Jawlik.
© 2016 John Wiley & Sons, Inc. Published 2016 by John Wiley & Sons, Inc.

Explanation

---
1. Errors in statistics can be classified into two kinds:
- **Experimental Errors:** someone or something did something wrong.
- **Sampling Errors:** a statistical estimate from a Sample is not identical to the property of the Population or Process which it estimates.

"Error" has the connotation of something gone wrong; a mistake has been made. And this is true in the kinds of errors classified as **Experimental Errors. A mistake was made; usually in the collection of data.**

However, Sampling "Error" is something of a misnomer. Nobody did anything wrong; no mistakes were made. **A Sampling "Error" might be better described as a Sampling "difference" or "deviation."** It is the difference between the numerical estimate of a property (a Statistic) calculated from a Sample (e.g., the Sample Mean) and the true value of that property (a Parameter) in the Population or Process from which the Sample was collected (e.g., the Population or Process Mean).

Usually, we do not know the true value of the Population or Process Parameter (otherwise we wouldn't be trying to estimate it with a Sample). So, the error is calculated from Probabilities.

This being statistics, there are several names for the same thing, and use of these names is not consistent. **Sampling Error is sometimes called "Random Error" or "Stochastic Error."**

---
2. Experimental Errors largely involve errors in collecting the data – Sampling Bias and Measurement Errors.

Sampling Bias occurs when a non-random Sample is collected. Some examples:
- Self-Selection Bias: e.g., a phone survey. The people surveyed are the ones who agreed to respond. These may be people who have a lot of time on their hands.
- Many social science surveys are performed on college campuses. Survey participants tend to be younger than the population at large.
- Inspectors of physical items tend to select for their Sample, those items which have a visible defect.

(For more, on this subject, see the article *Sample, Sampling*.)

Measurement Errors (also known as "Systematic Error" or "Observational Error") **can be caused by the measuring device, the person measuring, or both.** For example,

- The measuring device is faulty, or several inconsistently calibrated measuring devices are used.
- The needle on an analog meter rests between 2 and 3. The person reading the device does a visual interpolation in deciding whether to record a 2.6 or a 2.7.
- Different inspectors may make slightly different judgments on whether something constitutes a defect or not.

In quality improvement disciplines, like Six Sigma, there is an entire sub-discipline, Measurement System Analysis (MSA) devoted to this subject. Lack of Repeatability and Reproducibility are two major types of Measurement Errors examined in MSA.

 | **3. Sampling Errors include Alpha and Beta Errors, Margin of Error, Regression Residuals, Standard Error, and Sum of Squares Error.**

Sampling Errors are the most "statistical" of errors in statistics, and we'll devote most of this article to them. In fact much of Inferential Statistics is devoted to quantifying and studying these calculated differences between a Sample Statistic and the corresponding Population or Process Parameter.

Alpha Error (see the article *Alpha and Beta Errors*):

- **The error of falsely concluding that there is a difference, change, or effect, when there is not.** (That is, it is the error of rejecting the Null Hypothesis when it is true.)
- Also known as a **False Positive**

p, the p-value, is the Probability of an Alpha Error
Alpha (α) is the maximum value of p that we will accept in an Inferential Statistical analysis, such as a t-test or ANOVA.

Beta Error (see the article *Alpha and Beta Errors*):

- **The error of falsely concluding that there is not a difference, change, or effect, when there is.** (That is, it is the error of failing to reject the Null Hypothesis when it is false.)
- Also known as a **False Negative**

Margin of Error (see the article *Margin of Error*)

- symbol MOE or E
- **half the width of a 2-sided Confidence Interval**

For example, MOE is the 2% in the statement, "With a 95% Level of Confidence, and a Margin of Error of $\pm$ 2%, we predict that Candidate A will get 54% of the vote."

Random Error: another name for Sampling Error

Residuals in Regression (see the article *Residuals*)

For any value of the Independent Variable, x, in a Regression Model, the Residual is **the <u>difference</u> between the value of** the Dependent Variable, **y, predicted by the Model and the actual value of y** in the data. Design of Experiments (see the three articles on this subject) can help refine a Regression Model and reduce the size of the Residuals.

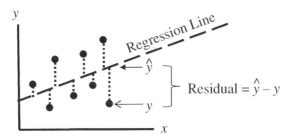

Standard Error: (see the article *Standard Error*).

Standard Error is defined as the Standard Deviation of the Sampling Distribution. It is frequently found as the denominator in Inferential Statistical formulas.

Stochastic Error: another name for Sampling Error

Sum of Squares Error (SSE): (see the article *Regression – Part 1: Sums of Squares*)

- the sum of the squared Residuals

- It **measures the Variation in y which is not explained by the Regression Model.**

Type I Error: another name for Alpha Error

Type II Error: another name for Beta Error

Some publications included Sampling Bias as a type of Sampling Error. According to our definition, it is classified as an Experimental Error.

 | **4. Some types of Sampling Errors influence each other.** |

Alpha Error and Beta Error

Alpha = 0.05 is the most common selection for the maximum Probability of an Alpha Error (False Positive) which we are willing to accept. One might ask, why not make it 99.9%? The reason is that the Probability of Alpha Error and Beta Errors affect each other inversely. If one goes down, the other goes up.

If you want to reduce the Probability of a False Positive (Alpha) in your conclusions, and you select a very small value for Alpha, you pay a price in the form of an increased Probability of Beta Error (False Negative). The article *Alpha and Beta Errors* has more on this.

Margin of Error (MOE) and Alpha Error

The Critical Value of a Test Statistic (such as z, t, F, or χ^2) is in the numerator of the formula for Margin of Error. So larger values for the Critical Value will result in a larger MOE. But, for any given Distribution, the Critical Value is dependent entirely on the value of Alpha. **A larger value for Alpha results in a smaller Critical Value.** And a smaller Critical Value is closer to the center of the Distribution. **So, a larger value for Alpha results in a smaller Margin of Error.**

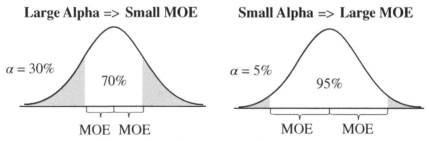

For more on this, see the article *Margin of Error*.

Standard Error

The Standard Error is used in Inferential Statistics to account for the error inherent in using a Sample for an estimate – the Sampling Error. It is a measure of how accurately a Sample represents a Population or Process.

It can be found in the denominator of the formulas for other Statistics. For example,

$$t = \frac{\textbf{sample Mean} - \textbf{specified Mean}}{\textbf{Standard Error}}$$

$$\text{Critical Value} = \frac{\textbf{Margin of Error}}{\textbf{Standard Error}}$$

The formula for Standard Error is different for different Statistics. Below are formulas for Standard Error of the Mean, of the difference between two Means, and of the Proportion.

$$\textbf{SEM} = \frac{s}{\sqrt{n}} \qquad \textbf{SE}\,(\bar{x}_1 - \bar{x}_2) = \sqrt{\frac{s_1^2}{n_1} + \frac{s_2^2}{n_2}} \qquad \textbf{SE}_p = \sqrt{\frac{p\,(1-p)}{n}}$$

For more, see the article *Standard Error*.

---

5. An increase in Sample Size reduces Sampling Errors.

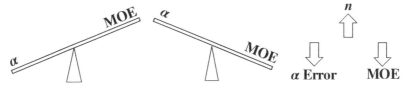

As we've shown, there are tradeoffs between different types of Sampling Errors. However, there is a way for us to "have our cake and eat it too." Increase the Sample Size, *n*, and you decrease Probability of a Sampling Error. For more, see the articles on *Sample Size*.

Related Articles in This Book: *Alpha and Beta Errors*; *Margin of Error*; *Regression – Part 1: Sums of Squares*; *Residuals*; *Sample Size – Parts 1 and 2*; *Sample, Sampling*; *Standard Error*; *Design of Experiments – Parts 1–3*

EXPONENTIAL DISTRIBUTION

Summary of Keys to Understanding

 1. **Exponential Distributions can be used to solve problems involving the time interval to or between events.**

 2. **Exponential Distributions can be used when events occur independently and at a constant average Rate.**

 3. **The Exponential is a family of Continuous data Distributions. An individual Distribution within the family can be specified by a single Parameter, either the Mean (μ) or the Rate (λ).**

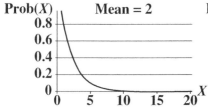 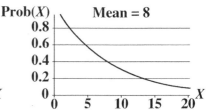

 4. **The Probability curve (PDF) of the Exponential Distribution approaches 1 at $X = 0$ and approaches 0 at $X =$ infinity.**

 5. **The Cumulative Probability (CDF) of the Distribution can be calculated with a formula.**

$$\text{Prob}(X > x) = e^{-x/\mu}$$

63% of this area under the curve is to the left of the Mean.

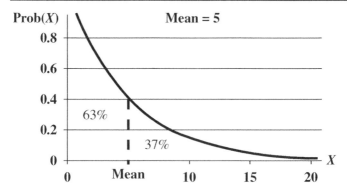

Explanation

 1. Exponential Distributions can be used to solve problems involving the time interval to or between events.

For example,

We own a small shop, and one new customer arrives, on average, about every 15 minutes. We would like to close up shop 10 minutes early today. What is the Probability that a new customer would arrive after we left but before our normal closing time?

The **Mean time interval** between customer arrivals is 15 minutes.

The Exponential Distribution can help us calculate the Probability of a customer arriving in the 10 minute interval between the time we leave early and our usual closing time.

Similar uses include:

– Time to complete a process step

– Time between failures

The Mean time between events is the inverse of the Rate at which the events occur.

In the example above, a Mean time between customer arrivals of 15 minutes corresponds to a Rate of 1/15 customers per minute. So, ...

$$\text{Mean} = 1/\text{Rate} \quad \text{and} \quad \text{Rate} = 1/\text{Mean}$$

The symbol for Mean is μ (the Greek letter mu), and the symbol for Rate is λ (the Greek letter lambda). So, ...

$$\mu = \frac{1}{\lambda}$$

and

$$\lambda = \frac{1}{\mu}$$

Terminology note: "Expected Value," E, is a term used almost synonymously with "Mean" in this context. In the above example, we started with a known value for the Mean interval and then calculated a value for the Rate. If we had, instead, started with a known value for the Rate, say four customers per hour, we could have then calculated an Expected Value for the time interval between future customers. That would be 15 minutes. We call it "expected" because it is a projection into the future, rather than a calculation of past data, like the Mean.

2. Exponential Distributions can be used when events occur independently and at a constant average Rate.

If a random customer comes into our store at a given time, that does not normally affect when another random customer comes in. The events occur independently. If, however, we had a barbershop with only one chair and no room for waiting, the next customer could not come in until the previous one was finished. So customer entries would not be independent.

A constant average Rate is a consequence of so-called "memoryless-ness" which is a property of Exponential Distributions. The Exponential Distributions works in "memoryless" situations, that is, when the past has no influence on the future.

For example, coin flipping is memoryless. The fact that you flipped heads 10 times in a row does not change the fact that the next coin flip has a 50% chance of being a head (or a tail).

However, memorylessness is not a property of things that decline with age. The time to engine breakdown for a car is not the same for a new car as for one with 150,000 miles. Car problems do not occur at a constant average Rate throughout the lifetime of a car.

 3. The Exponential is a family of Continuous data Distributions. An individual Distribution within the family can be specified by a single Parameter, either the Mean (μ) or the Rate (λ).

As described in the *Distributions – Part 1* article, **Continuous data Distributions are used with data that can have real number values** as opposed to integer-only values for Discrete data Distributions. And **they have smooth curves**, as opposed to the stairstep pattern of Discrete Distributions.

A Parameter is a numerical property of a Distribution ("Statistic" is the corresponding term for a property of a Sample.) Whereas it requires two Parameters – the Mean and the Standard Deviation – to describe a unique Normal Distribution, **only one Parameter is required to specify a unique member of the family of Exponential Distribution, either the Mean (μ) or the Rate (λ).**

Some texts will use the Mean to specify a unique Exponential Distribution, others use the Rate. Both contain the same information, so either can be used. For the two graphs below, the Rate is $\lambda = 1/2$ and 1/8, respectively.

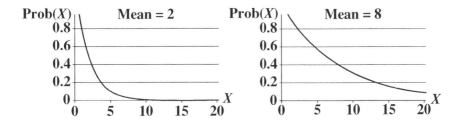

Prob(X) Mean = 2   Prob(X) Mean = 8

 4. The Probability curve (PDF) of the Exponential Distribution approaches 1 at x = 0 and approaches 0 at x = infinity.

Graphs of the Probability Density Function (PDF) of two members of the Exponential family of Distributions are shown above.

For $x < 0$, the Probability is 0.

For $x \geq 0$, Point Probability (PDF): **Prob**$(x) = \frac{1}{\mu}e^{-x/\mu} = \lambda e^{-\lambda x}$

where e is the Exponential Constant: $e = 2.718$.

In statistics, the Point Probabilities for individual values of x shown on the PDF are usually of little interest. The main use of a PDF is for calculating the Cumulative Probability of a range of values. For other Distributions, this is done by calculating the integral (remember that from calculus?) – the area under the PDF curve which is above the range of values on the horizontal axis. For Exponential distributions, this is not necessary.

 5. The Cumulative Probability (CDF) of the Distribution can be calculated with a formula.

$$\textbf{Prob}(X \leq x) = 1 - e^{-x/\mu}$$

63% of this area under the curve is to the left of the Mean; 37% is to the right.

Unlike most Distributions, the Cumulative Distribution Function (CDF) of the Exponential Distribution is a formula that does not involve the integral.

Cumulative Probability (CDF): **Prob**$(X \leq x) = 1 - e^{-x/\mu} = 1 - e^{-\lambda x}$

Where X is the horizontal axis Variable and x is a specified value of that Variable. In our example above, $x = 10$.

As shown in the graph below, most of the Cumulative Probability in any Exponential Distribution lies to the left of the Mean.

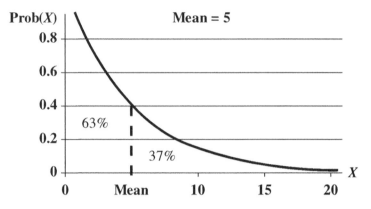

The split is about 63% to the left of the Mean and 37% to the right. This can be seen by plugging $x = \mu$ into the formula:

$$\mathbf{Prob}(X \leq \mu) = \mathbf{1} - e^{-\mu/\mu}$$
$$= 1 - 2.718^{-1}$$
$$= 1 - (1/2.718)$$
$$= 0.632 \sim \mathbf{63\%}$$
$$\mathbf{Prob}(X > \mu) = \mathbf{1} - \mathbf{Prob}(X \leq \mu)$$
$$= 0.368 \sim \mathbf{37\%}$$

<u>Back to our example</u> at the beginning of this article,

We own a small shop, and one new customer arrives, on average, about every 15 minutes ($\mu = 15$).

We would like to close up shop 10 minutes early today ($x = 10$).

What is the Probability that a new customer would arrive within the 10 minute interval between the time when we left but before our normal closing time? (Cumulative Probability of $x \leq 10$)

$$\mathbf{Prob}(X < x) = e^{-x/\mu}$$
$$= 2.718^{-10/15}$$
$$= (1/2.718)^{0.667}$$
$$= 51.3\%$$

Related Articles in This Book: *Distributions – Part 1: What They Are*; *Distributions – Part 2: How They Are Used*; *Distributions – Part 3: Which to Use When*

F

Summary of Keys to Understanding

 1. F is a Test Statistic which is the ratio of two Variances.

Samples: $F = (s_1)^2/(s_2)^2$ Populations: $F = (\sigma_1)^2/(\sigma_2)^2$

ANOVA: $F = \text{MSB}/\text{MSW}$

 2. F has a different Probability Distribution for each combination of Degrees of Freedom – for the numerator Sample and the denominator Sample.

   As the dfs grow larger, the Distributions become more symmetrical.

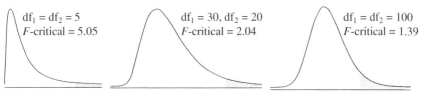

$\text{df}_1 = \text{df}_2 = 5$
F-critical = 5.05

$\text{df}_1 = 30, \text{df}_2 = 20$
F-critical = 2.04

$\text{df}_1 = \text{df}_2 = 100$
F-critical = 1.39

All F Distributions:

- **start at zero.** (F can never be negative.)
- **extend to the right to infinity** (and never touch the horizontal axis).
- **are <u>not</u> symmetrical.**
- **have a Median of roughly 1.**

 3. The F-test tests for "equal Variance" – that is, **whether there is a Statistically Significant difference in the Variation of two or more Populations or Processes.**

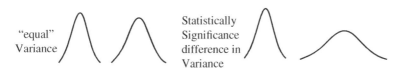

"equal" Variance

Statistically Significance difference in Variance

 4. The F-test is a key component of ANOVA.

Statistics from A to Z: Confusing Concepts Clarified, First Edition. Andrew A. Jawlik.
© 2016 John Wiley & Sons, Inc. Published 2016 by John Wiley & Sons, Inc.

Explanation

1. ***F* is a Test Statistic which is the ratio of two Variances.**

A **Test Statistic** is one whose **Distribution has known Probabilities.** So, for any value of *F* (on the horizontal axis below), there is a known Probability of that value occurring. That Probability is the height of the curve above that point.

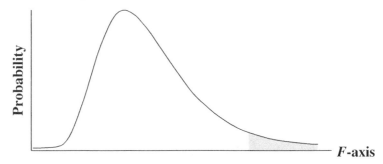

More importantly, we can calculate the area under the curve beyond the value of *F*. This gives us a Cumulative Probability (such *p*) which we can use to compare to the Significance Level, *α* (also a Cumulative Probability), in various types of analyses in Inferential Statistics. A Cumulative Probability is usually depicted as a shaded area under the curve.

***F* is the ratio of Two Variances**. To keep things simple, **the larger Variance is entered as the numerator, and the smaller is the denominator**, except for ANOVA, where the numerator and denominator are specified:

- **For Samples: $F = (s_1)^2/(s_2)^2$**

where s_1 **and** s_2 is the symbol for the Standard Deviations of a Samples 1 and 2, and their squares are the Variances.

- **For Populations or Processes: $F = (\sigma_1)^2/(\sigma_2)^2$**

σ_1 and σ_2 are the Standard Deviations of Populations 1 and 2, respectively. So, $(\sigma_1)^2$ and $(\sigma_2)^2$ are the respective Variances.

- **In ANOVA: $F = MSB/MSW$,**

where MSB is the Mean Sum of Squares Between and MSW is the Mean Sum of Squares Within. Both are special types of Variances. See the *ANOVA – Part 2* article.

F is Continuous Distribution. Its curve has a smooth shape, unlike the histogram-like shape of Discrete Data Distributions. However, it can work with Samples of Discrete data. The ratio of the Variances of two sets of Discrete Data is Continuous.

> **2. *F* has a different Probability Distribution for each combination of the two Degrees of Freedom** – for the numerator Sample and the denominator Sample.
>
>   **As the dfs grow larger, the Distributions become more symmetrical.**

(The shaded area $\alpha = 0.05$ in these diagrams)

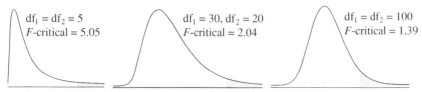

$df_1 = df_2 = 5$
F-critical = 5.05

$df_1 = 30, df_2 = 20$
F-critical = 2.04

$df_1 = df_2 = 100$
F-critical = 1.39

All *F* Distributions:

- **start at zero.** (*F* can never be negative.)
- **extend to the right to Infinity** (and never touch the horizontal axis).
- **are <u>not</u> symmetrical.**
- **have a Median of roughly 1.**

The *F*-statistic has a different Distribution for each combination of the two values of Degrees of Freedom, df. For *F*, df $= n - 1$, where *n* is the Sample Size. Since *F* is the ratio of two Variances, and since each one has their own df, *F* has a different Distribution for each combination of dfs.

F is never negative, because Variances, being squares, cannot be negative.

Tables of Critical Values of *F* have the columns represent one of the dfs and the rows the other df. The third Variable in calculating *F*-critical is the Significance Level, α. Since we can't have 3-dimensional tables, **there is usually a different table of Critical Values for each value of α.**

> **3. The *F*-test tests for "equal Variance"** – that is, **whether there is a Statistically Significant difference in the Variation of two or more Populations or Processes.**

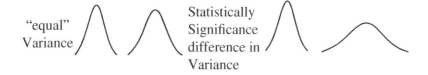

"equal" Variance

Statistically Significance difference in Variance

A number of statistical tests, including the *t*-tests, **assume equal Variance** between the two Populations or Processes that are being analyzed. "Equal" does not mean identical. It means that the Variances are close enough that there is no Statistically Significant difference between them. **The *F*-test can be used to determine this.**

Also **when one is trying to determine whether two Populations or Processes are statistically the same or different, Variation** (of which Variance is a good measure) **is one of three criteria which can be evaluated.** The other two are Central Tendency (e.g., the Mean) and Shape (Skewness and Kurtosis).

As we noted earlier, the *F* for two Samples is:

$$F = (s_1)^2/(s_2)^2$$

where s^2 is the symbol for a Sample Variance. (*s* is the symbol for Sample Standard Deviation.)

If *F* is close enough to 1, then there is not a Statistically Significant difference between the Variances of the two Samples.

How do we know if *F* is close enough to 1? The *F*-test does an Inferential Statistical analysis involving the Alpha we select and the *F* Distribution with dfs of n_1 and n_2. It produces a *p*-value which we can compare with Alpha. It also produces values for *F* and *F*-critical, which we can compare with each other. Either of the following two comparisons can be used. They are equivalent:

If $p \leq \alpha$, or, equivalently if $F \geq F$-critical, then
> there is a Statistically Significant difference – Reject the Null Hypothesis)

Otherwise,
> there is not a Statistically Significant difference – Fail to Reject (i.e., Accept) H_0

Assumptions: **The *F*-test requires that**

- **the Sample data be approximately Normal**
- **the two Samples are independent**

If the data are not approximately Normal, Levene's test can be used instead.

"Independent" means that there is not a Correlation or Association between the two Samples. For example, "before" and "after" Samples of the same test subjects would not be independent.

Outliers: In addition, **the accuracy of the test can be adversely affected by Outliers** especially in small Samples. Outliers can have an outsize effect on the Variance.

Another test for the Variance: Whereas the F-test compares the Variances of two Populations or Processes, **the Chi-Square Test for the Variance compares the Variance of one Population or Process to a Variance we specify.** A typical specified Variance could be an estimate, a target, or a historical value.

(If you're familiar with t-tests, the F-test would be analogous to the 2-Sample t-test, and the Chi-Square Test for the Variance would be analogous to the 1-Sample t-test.)

4. The F-test is a key component of ANOVA.

ANOVA is a type of Inferential Statistical analysis. It is an acronym for "Analysis of Variance." This is something of a misnomer. **The purpose of ANOVA is to analyze several Sample Means** (usually three or more) to determine whether there is a Statistically Significant difference among them. **The method by which it does this is by analyzing Variances.**

t-tests are usually used to analyze differences between two Means. If we have three Means, we could do three pairwise comparisons, two Means at a time. However, each of those t-tests would bring an Alpha Error (the error of a False Positive). And that error would be compounded three times.

However – as explained in the *ANOVA – Part 2* article – **ANOVA can take information on multiple Means and distill them into two Statistics.** These are the Mean Sum of Squares Between (MSB) and the Mean Sum of Squares Within (MSW). **MSB and MSW are two special kinds of Variances.** Then, ANOVA uses these two Statistics in a single F-test to accomplish the same thing that multiple t-tests would accomplish. This avoids compounding the Alpha Errors.

$$F = \frac{\text{MSB}}{\text{MSW}}$$

If the value of F is close to 1, then there is no Statistically Significant difference between the Means of the Populations or Processes being compared.

The output of an ANOVA is a table like the one below. An ANOVA table is also provided in Regression analyses.

ANOVA	**Cannot Reject Null Hypothesis because $p > 0.05$ (Means are the same)**					
Source of Variation	**SS**	**df**	**MS**	**F**	**p-value**	**F-crit**
Between Groups	44970.83	9	4996.759	1.840992	0.123	2.39814
Within Groups	54283.33	20	2714.167			
Total	99264.17	29				

In this example, we see that $F < F$-critical (and, equivalently, $p > \alpha$). So we Fail to Reject (we Accept) the Null Hypothesis. That is, we conclude that there is <u>not</u> a Statistically Significant difference in the Variances MSB and MSW. So, there is <u>not</u> a Statistically Significant difference among the Means in this ANOVA.

Related Articles in This Book: *Test Statistic; Variance; Distributions – Parts 1–3; Degrees of Freedom; Inferential Statistics; ANOVA – Parts 1 and 2; Alpha (α); p-Value, p; Critical Value; Chi-Square Test for the Variance*

FAIL TO REJECT THE NULL HYPOTHESIS

Summary of Keys to Understanding

 1. **"Fail to Reject the Null Hypothesis" is 1 of 2 possible conclusions from a Hypothesis Test.** The other conclusion is "Reject the Null Hypothesis."

   If $p > \alpha$, Fail to Reject the Null Hypothesis.

 2. **A Null Hypothesis states that there is no** (Statistically Significant) **difference, change, or effect. "Fail" and "Reject" cancel each other out, leaving the Null Hypothesis in place** when we fail to reject it.

A statistician responds to a marriage proposal:

 3. **Practically speaking, it is OK to "Accept the Null Hypothesis." You don't have to "Fail to Reject."**

Explanation

It may be helpful to read the articles "Null Hypothesis" and "Reject the Null Hypothesis" before reading this one.

 1. "Fail to Reject the Null Hypothesis" is 1 of 2 possible conclusions from a Hypothesis Test. The other conclusion is "Reject the Null Hypothesis."

If $p > \alpha$, Fail to Reject the Null Hypothesis.

As stated in the article in this book, *"Hypothesis Testing – Part 1: Overview,"* Hypothesis Testing is one method of Inferential Statistics. It is a method for answering questions about a Population or a Process by analyzing data from a Sample.

The *"Part 2"* article describes the 5 steps in this method. In Step 1, we select a value for the Level of Significance, Alpha (α). In Step 4, the analysis calculates a value for p, the Probability of an Alpha Error.

Step 5 is to come to a conclusion about the Null Hypothesis by comparing p to Alpha. There are only two possible conclusions:

- **If $p > \alpha$, Fail to Reject the Null Hypothesis**
- **If $p \leq \alpha$, Reject the Null Hypothesis**

Inferential Statistical Analyses, such as t-tests or ANOVA, also calculate the value of a Test Statistic – for example, t – and the Critical Value of the Test Statistic – t-critical. Comparing t to t-critical is the same as comparing p to Alpha. You can do either one; they are redundant. But the "$\leq$" is in the opposite direction.

- **If $t < t$-critical, Fail to Reject the Null Hypothesis**
- **If $t \geq t$-critical, Reject the Null Hypothesis**

See the article in this book, *p, t, and F: "$>$" or "$<$" ?* for some tips on how to remember which way the comparison symbol points. It may also be helpful to read the article, *"Alpha, p-value, Critical Value, and Test Statistic – How They Work Together."*

 2. A Null Hypothesis states that there is no (Statistically Significant) **difference, change, or effect. "Fail" and "Reject" cancel each other out, leaving the Null Hypothesis in place** when we fail to reject it.

Examples of Null Hypotheses:

- There is <u>no difference</u> between the Mean effectiveness of the two medicines.
- There has been <u>no change</u> in the Process Standard Deviation from its historical value.
- The training program has had <u>no effect</u> on worker performance.

<u>Triple Negatives:</u>

I <u>fail</u>
to <u>reject</u>
the <u>Null</u>
Hypothesis.

I <u>don't</u>
<u>not</u> have
<u>no</u>
money.

We are all taught in elementary school to avoid using double negatives like, "I <u>don't</u> have <u>no</u> money." However, statistics goes beyond the double negative, to an even-more-confusing triple negative: "<u>Fail</u> to <u>Reject</u> the <u>Null</u> Hypothesis."

"Fail" and "Reject" cancel each other out, leaving the Null Hypothesis standing. Perhaps the following table may help make this clearer.

Positive statement	There is a difference between the two Means.	$+1$
Null Hypothesis (H_0)	There is <u>no</u> difference between the two Means.	-1
Reject H_0	There is <u>not no</u> difference … i.e., There is a difference …	$-1 \times -1 = +1 =$ Positive statement
Fail to Reject H_0	There is <u>not not no</u> difference … i.e., There is no difference	$-1 \times -1 \times -1 = -1$ = Null Hypothesis

 3. Practically speaking, it is OK to "Accept the Null Hypothesis." You don't have to "Fail to Reject."

So, if you don't reject H_0, then you accept it – right? Many experts insist on saying "Fail to Reject" because the Null Hypothesis is a negative (no difference, no change, no effect), and one can't prove a negative.

"You can't prove a negative." is true – <u>if</u> you require 100% accuracy. But, Hypothesis Testing does <u>not</u> strive for 100% accuracy.

In fact, if we <u>could</u> get 100% accuracy, we would not need Inferential Statistics. We would just answer our question via counting, measurement, or a precise formula.

An integral part of Hypothesis Testing is accepting a specified Probability of error. Before collecting data, we must select a Level of Significance, Alpha (α), which is the Probability of a "False Positive." Most often, $\alpha = 5\%$ is selected. So, we know we're not getting 100% accuracy.

<u>One consequence of insisting that "You can't prove a negative":</u>

You can't prove that I don't exist and that I'm not hiding out in Chicago.

So, some say that the most you can do is "Fail to Reject" the Null Hypothesis that there are no unicorns living in Chicago.

Also, practically speaking, we're not going to conduct an experiment or test, spend significant time and money on it, and then go tell our boss or our customer that we can't conclude anything from it.

So, **we may <u>not</u> want to say that we have <u>proven</u> the Null Hypothesis. But we <u>can</u> proceed to <u>take action as if</u> we Accept the Null Hypothesis.** And many experts say that it's fine to come out and say that we Accept the Null Hypothesis.

<u>Parsing the Statisticians' Response to the Marriage Proposal:</u>

(accept)   *(no change in our status)*

I Fail to Reject the Null Hypothesis.

Will you marry me?

Oh No! That means "No"! The Null Hypothesis is a negative, and to Fail to Reject the negative leaves it in place.

What is the <u>Null Hypothesis</u> in the case of the Statisticians Marriage Proposal?

- Prior to the proposal of marriage, they were <u>not</u> engaged to be married.
- A Null Hypothesis means there is no change, no effect, or no difference.
- So the **Null Hypothesis would be that things remain the same for this couple. There is <u>no change</u> in their status – no new status of being engaged to be married.**
- And, **by Failing to Reject it, she accepts the Null Hypothesis.** That is, she accepts no change. **They will <u>not</u> be engaged.**

Related Articles in This book: *Null Hypothesis*; *Reject the Null Hypothesis*; *Hypothesis Testing – Part 1: Overview*; *Hypothesis Testing – Part 2: How To*; *Null Hypothesis*; *Alternative Hypothesis*; *p, t, and F: ">" or "<"?*; *Alpha, p-Value, Critical Value, and Test Statistic – How They Work Together*

HYPERGEOMETRIC DISTRIBUTION

 | **1. The Hypergeometric is a Distribution for Discrete data. Units are counted, not Occurrences.**

Discrete data are integers, such as Counts. Counts are non-negative. In contrast to Continuous data, there are no intermediate values between consecutive integer values.

The Hypergeometric Distribution is used for Counts of Units, such as the number of shirts manufactured with defects. Units are different from Occurrences. If a shirt (the Unit) we inspected had three defects, we would add only one to the Count of defective Units, and we could use the Hypergeometric Distribution.

If we were interested in the total number of Occurrences of defects – not the number of defective Units – we would count three Occurrences for that shirt and we would use a different Discrete data Distribution – the Poisson Distribution.

 | **2. Use the Hypergeometric Distribution**, instead of the Binomial Distribution, **when**
- **Sampling <u>without</u> Replacement**
- **the Sample Size is large** relative to the Population, i.e., when $n > 10\%$ of N

To illustrate the concept of Replacement, let's say we're doing a study in a small lake to determine the Proportion of Lake Trout. If we throw the fish back in before trying to catch the next fish that is called Sampling <u>with</u> Replacement. When Sampling with Replacement, a different Discrete data Distribution should be used – the Binomial Distribution.

But, if we keep the fish, that is **Sampling <u>without</u> Replacement**, and **the Hypergeometric Distribution should be used.**

We should also use the Hypergeometric, instead of the Binomial, **when the Sample Size (n) is large, relative to the Population size (N).** If the Sample Size is more than 10% of the Population size, use the Hypergeometric.

Statistics from A to Z: Confusing Concepts Clarified, First Edition. Andrew A. Jawlik.
© 2016 John Wiley & Sons, Inc. Published 2016 by John Wiley & Sons, Inc.

3. The Probability of exactly X units in a Sample is **a function of N, the Population Size, n, the Sample Size, and D, the number of counted Units (e.g., defective shirts) in the Population.**

This can be calculated using spreadsheets or software.

Related Articles in This Book: *Binomial Distribution*; *Distributions —* *Parts 1–3*; *Poisson Distribution*

HYPOTHESIS TESTING – PART 1 (OF 2): OVERVIEW

Summary of Keys to Understanding

 1. Hypothesis Testing is one method of Inferential Statistics, that is, for answering a question about a Population or Process, based on analysis of data from a Sample.

 2. The question usually asks whether there is a Statistically Significant difference, change, or effect. The question is converted to a negative statement called a Null Hypothesis (symbol H_0).

 3. There are two possible outcomes from a Hypothesis Test:
   • **Reject the Null Hypothesis, or**
   • **Fail to Reject the Null Hypothesis**

Question	Equivalent **Null Hypothesis**: (Negative Statement)	**Answer to the Question if we Reject** the Null Hypothesis	**Answer to the Question if we Fail to Reject** the Null Hypothesis
Q: **Is there** a Statistically Significant difference between the Means of these two Populations?	**There is no** difference in the Means of these two Populations.	Yes.	No.

 4. For some types of tests, you can use the Confidence Intervals method instead of Hypothesis Testing.

Explanation

Hypothesis Testing can be confusing for many people. So, this book devotes a fair amount of space to explaining it in several ways and in bite-sized chunks.

The core concept of Hypothesis Testing is addressed by two articles in this book:

- *Hypothesis Testing – Part 1: Overview* (this article)
- *Hypothesis Testing – Part 2: How To*

That information is expanded upon in four other articles which are essential to understanding the concept:

- *Null Hypothesis*
- *Alternative Hypothesis*
- *Reject the Null Hypothesis*
- *Fail to Reject the Null Hypothesis*

In addition, there are separate articles on related concepts mentioned in these articles, such as *Alpha (α)*, *Alpha and Beta Errors*, *p-Value*, *Critical Value*, *Confidence Interval*, *Test Statistic*, etc.

> **1. Hypothesis Testing is one method of Inferential Statistics,** that is, for answering a question about a Population or Process, based on an analysis of data from a Sample.

In Descriptive Statistics, we have complete data on the entire universe we wish to observe. So we can just calculate various properties (Parameters) of the Population or Process.

On the other hand, **in Inferential Statistics** (aka "Statistical Inference"), **we don't have the complete data**. The Population or Process is too big, or it is always changing. So we can never be 100% sure about it. **We can collect a Sample of data and make <u>estimates</u> of the Population or Process Parameters** (such as the Mean or Standard Deviation).

Hypothesis Testing and Confidence Intervals are the two main methods for doing this.

> **2. The question usually asks whether there is a Statistically Significant <u>difference</u>, <u>change</u>, or <u>effect</u>. The question is converted to a negative statement called a Null Hypothesis** (symbol H_0).

For example,

- Is there a Statistically Significant <u>difference</u> in the Means of these two Populations?
- Is there a Statistically Significant <u>difference</u> between the actual Mean lifetime of our light bulbs and the target Mean?
- Has there been a Statistically Significant <u>change</u> in the Standard Deviation of our Process from its historical value?
- Does this experimental medical treatment have a Statistically Significant <u>effect</u>?

To use the Hypothesis Testing method, we first need to convert the Inferential Statistics question into a negative statement, the Null Hypothesis:

Question	Equivalent Null Hypothesis (H_0) (Negative Statement)
Q: **Is there** a Statistically Significant <u>difference</u> between the Means of these two Populations?	H_0: **There is no** <u>difference</u> in the Means of these two Populations.
Q: **Is there** a Statistically Significant <u>difference</u> between the actual Mean lifetime of our light bulbs and the target Mean?	H_0: **There is no** <u>difference</u> between the actual Mean lifetime of our light bulbs and the target Mean.
Q: **Has there been** a Statistically Significant <u>change</u> in the Standard Deviation of our Process from its historical value?	H_0: **There has been no** <u>change</u> in the Standard Deviation of our Process from its historical value.
Q: **Does this** experimental medical treatment **have** a Statistically Significant <u>effect</u>?	H_0: **This** experimental medical treatment **does not have** an <u>effect</u>.

If you understand this so far, you've overcome half the confusion which people have with Hypothesis Testing. (The other half comes from the two possible outcomes of Hypothesis testing: "Reject the Null Hypothesis" or "Fail to Reject the Null Hypothesis.") But even if you're still confused, there is another chance. Please read the article *Null Hypothesis* in this book, and hopefully that will help clear things up.

> **3. There are two possible outcomes from a Hypothesis Test:**
> - **Reject the Null Hypothesis, or**
> - **Fail to Reject the Null Hypothesis**

These are very confusing for most people. Since the Null Hypothesis is a negative statement, **Reject the Null Hypothesis is a double negative**. This is akin to saying, "I don't have no money." And we're all taught in elementary school to avoid talking like that.

"I don't have no money," the double negative, actually means "I <u>do</u> have money." So the double negative becomes a positive.

Let's go back to one of our Inferential Statistical questions:

Question	Equivalent **Null Hypothesis** (Negative Statement)	**Answer to the Question if we Reject** the Null Hypothesis
Q: **Is there** a Statistically Significant <u>difference</u> between the Means of these two Populations?	**There is no** <u>difference</u> in the Means of these two Populations.	**Yes.**

Fail to Reject the Null Hypothesis is even worse; it **is a triple negative.** Triple Negatives:

I <u>fail</u>
to <u>reject</u>
the <u>Null</u>
Hypothesis.

I <u>don't</u>
<u>not</u> have
<u>no</u>
money.

Question	Equivalent **Null Hypothesis** (Negative Statement)	**Answer to the Question if we Reject** the Null Hypothesis	**Answer to the Question if we Fail to Reject** the Null Hypothesis
Q: **Is there** a Statistically Significant <u>difference</u> between the Means of these two Populations?	**There is no** <u>difference</u> in the Means of these two Populations.	**Yes.**	**No.**

There is a separate article on Reject the Null Hypothesis and another on Fail to Reject the Null Hypothesis. These offer different ways of explaining what we just covered, in case it's helpful.

| **4. For some types of tests, you can use the Confidence Intervals method instead** of Hypothesis Testing.

Hypothesis Testing has some drawbacks:

- **The language can be confusing, as we have shown.**
- **It can appear to be inconclusive.**

Many authorities are adamant that we must "Fail to Reject" the Null Hypothesis and that we cannot Accept it. This seems to indicate that the results were inconclusive.

- **Experts disagree on key concepts in Hypothesis Testing.**

"Fail to Reject" is one of these. Some experts have a contrary view to that stated above. They say that the Null Hypothesis is the default condition, so that if we fail to Reject it, the Null Hypothesis stands. Similarly, others say – and this book agrees – that one can just proceed as if one Accepts the Null Hypothesis.

Also, there is strong disagreement on whether or not an Alternative Hypothesis must be stated. The article in this book on that subject explains how an Alternative Hypothesis can be helpful in certain situations (i.e., 1-tailed tests).

For these reasons, many people prefer a different way. The Confidence Interval method is the other main method of Inferential Statistics. Some experts recommend using it instead of Hypothesis Testing whenever possible. The Confidence Interval method is less confusing, and Confidence Intervals are more graphical.

- **The Confidence Interval method can be used when comparing the Parameter (e.g., the Mean) of <u>one</u> Population or Process to a specified Parameter** (e.g., a target we specify, a historical value, or zero).
- **However, when comparing Parameters of <u>two</u> Populations or Processes, it can give unclear results,** which may require a Hypothesis Test to resolve. (*See the article Confidence Intervals – Part 2: Some Specifics.*)

Examples:

Null Hypothesis (H_0) Negative Statement	Specified Parameter	Use
H_0: **There is no** difference between the actual Mean lifetime of our light bulbs and the <u>target</u> Mean.	target Mean	Confidence Intervals Hypothesis Testing
H_0: **There is no** difference in the Means of these <u>two</u> Populations.	not applicable	Hypothesis Testing

Related Articles in This Book: *Hypothesis Testing – Part 2: How to*; *Null Hypothesis*; *Alternative Hypothesis*; *Reject the Null Hypothesis*; *Fail to Reject the Null Hypothesis*; *Confidence Intervals – Parts 1 and 2*

HYPOTHESIS TESTING – PART 2 (OF 2): HOW TO

Summary of Keys to Understanding

> ### Five–Step Method for Hypothesis Testing
>
> 1. **State the problem or question in the form of a Null Hypothesis (H_0) and Alternative Hypothesis (H_A).**
> 2. **Select a Level of Significance (α),**
>   *This is the maximum level of risk for a "False Positive," i.e., an Alpha Error that we are willing to tolerate. Most frequently, it's 0.05.*
> 3. **Collect a Sample of data for analysis.**
> 4. **Perform a statistical analysis on the Sample data.**
>   *For example, t-test, F-test, ANOVA*
> 5. **Come to a Conclusion about the Null Hypothesis.**
>   *Reject it or Fail to Reject it.*

Example of Hypothesis Testing viewed as Input/Processing/Output

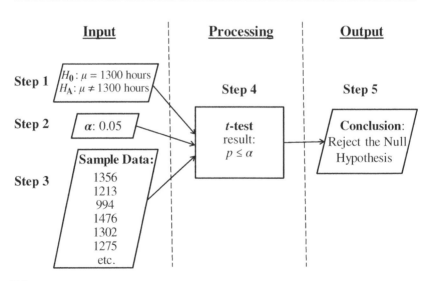

Explanation

In Steps 1, 2, and 3, we provide the inputs.

Step 1: **State the problem or question in the form of a Null Hypothesis (H_0) and Alternative Hypothesis (H_A).**

The Null Hypothesis (H_0) is a statement that there is **no Statistically Significant**

- **difference,**
- **change,** or
- **effect.**

For example, if we want to know whether the current Mean lifetime of the light bulbs we make has changed from the historical Mean of 1300 hours, our Null Hypothesis would be:

"There is no Statistically Significant difference between the current Mean lifetime of our light bulbs and the historical Mean of 1300 hours."

Or, more simply, "$\mu = 1300$," where μ is the symbol for the current Mean.

There is some disagreement among experts whether an Alternative Hypothesis (H_A) is necessary or desirable. It is included in this method of Hypothesis Testing, because it helps make obvious whether the test is 2-tailed, left-tailed, or right-tailed. And we need to know that in performing Step 3 of this method.

Recommendation: please read the articles *Null Hypothesis* and *Alternative Hypothesis* – in that order.

Step 2: **Select a Level of Significance, Alpha (α).**

This is where we define what is Statistically Significant or not. The concept of Alpha is covered extensively in this book in the article "*Alpha, α*" and related articles referenced there. But briefly, Alpha is the maximum Probability of a "False Positive" Error (aka "Alpha Error" or "Type I Error") which we are willing to tolerate and still call our results Statistically Significant.

Alpha thus is a clip level, a maximum tolerable level for the Probability, p, of an Alpha Error. If value of p (the "p-value), which is calculated in Step 4, is less than or equal to Alpha (i.e., $p \leq \alpha$), then any observed difference, change, or effect is deemed to be Statistically Significant.

See the article *Alpha, α*.

Step 3: **Collect a Sample of data.**

It is essential to the integrity of the Hypothesis Test that Steps 1 and 2 be completed before collecting the Sample of data. We don't want our framing of the Null Hypothesis or our selection of the value of Alpha to be tainted by how the data may appear to us.

Input

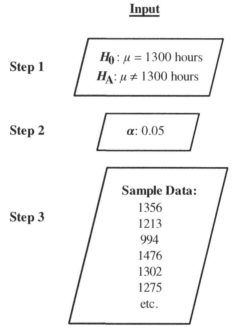

Step 1

$H_0 : \mu = 1300$ hours
$H_A : \mu \neq 1300$ hours

Step 2

α: 0.05

Step 3

Sample Data:
1356
1213
994
1476
1302
1275
etc.

Steps 1–3 give us the input portion of our Input/Processing/Output way of looking at our five-step method for Hypothesis Testing.

Step 4: Perform the Statistical Analysis.

This can be something like a *t*-test, *F*-test, Chi-Square tests, ANOVA, etc. (See the article *Which Statistical Test to Use to Solve Some Common Problems*.) Since our example uses the Mean of a single Population or Process in the Null Hypothesis, we will use a 1-Sample *t*-test. In this test, *t* is the Test Statistic.

The analysis will use the Sample data to:

- calculate a value for the Test Statistic, *t*
- calculate the *p*-value

It will also use our selected value of Alpha and the Probability Distribution of the Test Statistic to:

- calculate the Critical Value of the Test Statistic, *t*-critical

We then can make either one of two comparisons; these are statistically identical:

- Compare the *p*-value to Alpha
- Compare the calculated value of the Test Statistic, *t*, to its Critical Value, *t*-critical

Step 5: Come to a Conclusion about the Null Hypothesis, based on the results of the analysis. We either

- Reject the Null Hypothesis if
  - $p \leq \alpha$
    (which is statistically identical to...)
  - the calculated value of the Test Statistic $\geq$ Critical Value (e.g., $t \geq t$-critical)
- or Fail to Reject the Null Hypothesis if the results are the opposite, that is, if
  - $p > \alpha$
    (which is statistically identical to...)
- the calculated value of the Test Statistic $<$ Critical Value (e.g., $t < t$-critical)

There are separate articles in this book on each of these two conclusions.

Related Articles in This Book: *Null Hypothesis*; *Alternative Hypothesis*; *Alpha, α*; *p-Value, p*; *Alpha, p-Value, Critical Value, and Test Statistic – How they Work Together*; *p, t, and F: ">" or "<" ?*; *Alpha and Beta Errors*; *Reject the Null Hypothesis*; *Fail to Reject the Null Hypothesis*; *Hypothesis Testing – Part 1: Overview*

INFERENTIAL STATISTICS

Inferential Statistics (aka Statistical Inference) **is a huge umbrella topic whose scope includes many of the articles comprising this book.** This article is a high-level overview. For a similarly wide-scope article addressing Inferential Statistical tools, see the article *Which Statistical Tool to Use to Solve Some Common Problems.*

Summary of Keys to Understanding

 1. **In Inferential Statistics, we calculate a numerical property of a Sample** of data (e.g., the Sample Mean) **and use it to infer** (estimate) **the value of that property for the Population or Process from which the Sample was collected.**

 2. **We make these inferences with a specified Level of Confidence. The level of Confidence is 1 minus the Level of Significance, Alpha (α).** We select the value for Alpha.

I want a 95% Level of Confidence   Then, you should select $\alpha = 5\%$

 3. **Confidence Intervals is one of the two main methods used in Inferential Statistics.**

 4. **Hypothesis Testing is the other.**

 5. **Inferential Statistics is involved in such analyses as ANOM, ANOVA, Chi-Square Tests, *F*-tests, Regression, *t*-tests, and *z*-tests.**

Statistics from A to Z: Confusing Concepts Clarified, First Edition. Andrew A. Jawlik.
© 2016 John Wiley & Sons, Inc. Published 2016 by John Wiley & Sons, Inc.

Explanation

 1. **In Inferential Statistics, we calculate a numerical property of a Sample** of data (e.g., the Sample Mean) **and use it to infer** (estimate) **the value of that property for the Population or Process from which the Sample was collected.**

In Descriptive Statistics, we have all the data on an entire Population or Process. So, we can calculate numerical values which describe its statistical properties. For a Population or Process, these are properties called Parameters. Examples of Parameters include Mean, Mode, Median, Proportion, Standard Deviation, Variance, Skewness, and Kurtosis.

But **most often, we don't have all the data from the entire universe under consideration.** It is impractical or impossible to collect data from all the residents of a country, for example. And, for an ongoing Process, new data are constantly being created, so whatever we gather will be incomplete soon afterward.

So, **we collect a Sample of data, and we calculate a statistical property for the Sample.** This is called a Statistic. For every Population or process Parameter, there is a corresponding Sample Statistic. **In Inferential Statistics, the value of the Sample Statistic,** e.g., the Sample Mean, **becomes our estimate** (inference) **of the value of its corresponding Population or Process Parameter,** e.g., the Population Mean.

But how good is this estimate?

Since it's usually impossible to know the exact value of a Population or Process Parameter, one might think that we could never know with 100% accuracy if our Sample's estimate is accurate or even close. But through Inferential Statistics, we can specify precisely the Level of Confidence that we need.

For example, let's say we measured the height of a Sample of adult males and calculated the Mean (average) as 175 cm. We can specify that we want a 95% Level of Confidence. Depending on the Sample Size and the amount of Variation of the measurements in the Sample, we could come up with a result that said,

- With a **95% Level of Confidence**
- **the Mean** height of adult males in the Population **is 175 cm**
- **plus or minus 5 cm**

The "plus or minus 5 cm" tells us this is an Interval Estimate. More on this, when we get to Confidence Intervals.

| **2. We make these inferences with a specified Level of Confidence. The level of Confidence is 1 minus the Level of Significance, Alpha (α).** We select the value for Alpha.

We can never be 100% certain about an estimate. But Inferential Statistics enables us to get close to that. In fact the most common Level of Confidence is 95% (0.95). **The good news is that, in Inferential Statistics, we get to select the level of Confidence.**

But what exactly are we confident about when we say we want a certain Level of Confidence? We want to be confident that our conclusion is not a False Positive. A False Positive would be to conclude that the Mean height is 175 cm $\pm$ 5 cm, when in fact the true Population Mean was outside that range.

A False Positive is called an Alpha Error or Type I Error. Alpha (α) is the maximum Probability of an Alpha Error which we are willing to tolerate. We get to select the value of Alpha.

$$\alpha = 1 - \text{Level of Confidence}$$

If we want a 95% Level of Confidence, we select Alpha = 5%. (Most spreadsheets or software will expect you enter this as 0.05.)

I want a 95% Level of Confidence   Then, you should select $\alpha = 5\%$

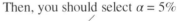

The concept of Alpha is central to understanding Inferential Statistics, so you may want to read the article *Alpha, α*. There is a separate article on *Alpha and Beta Errors*.

 | **3. Confidence Intervals is one of the two main methods used in Inferential Statistics.**

Earlier, we gave an example of an Interval Estimate: the Mean height of adult males in the Population is 175 cm plus or minus 5 cm. (This "plus or minus" amount, 5 cm, is called the *Margin of Error*; see the article by that name.) Our Inferential Statistical analysis had come to the conclusion that the true Population Mean is somewhere in the interval of 170–180 cm, with a 95% Level of Confidence. That interval is a Confidence Interval.

We use the selected value of Alpha to mark off shaded areas under one or both tails of the Distribution curve of a Test Statistic such as z. In this example, a "plus or minus" interval tells us to split Alpha (5%) in half and shade, under each tail, 2.5% of the area under the curve.

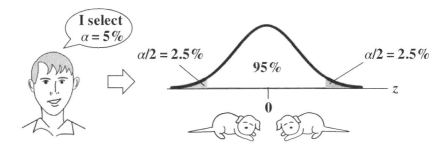

The inner boundaries of these shaded areas are the Critical Values of the Test Statistic. Between these boundaries is an interval. But it is in the units of the Test Statistic (z, in this example). We need to convert it to units of the original data (centimeters) in order to get the Confidence Limits which define the Confidence Interval.

We just use the formula for the Test Statistic to make this conversion as shown below.

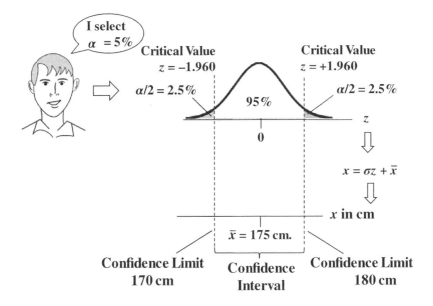

The Confidence Interval method can be used to determine whether there is a Statistically Significant difference between our Sample Statistic and a specified value for a Parameter (say a historical Mean or a target Mean for a Population or Process). In this type of analysis, **if the specified value is within the Confidence Interval, we conclude that there is <u>not</u> a Statistically Significant Difference** between that specified value and the actual value for the Population or Process.

The Confidence Intervals (CI) method is less useful for analyses in which Parameters from <u>two</u> Populations or Processes are compared. **If their CIs do <u>not</u> overlap, we conclude that there <u>is</u> a Statistically Significant difference between them**. **However, the converse is not always true.** If the CIs do overlap somewhat, we <u>cannot</u> conclude that there is <u>not</u> a Statistically Significant difference. We would need to use a Hypothesis Test to be certain. For more on this, see the article, *Confidence Intervals – Part 2: Some Specifics*.

 | **4. Hypothesis Testing is the other of the two main methods used in Inferential Statistics.**

Hypothesis Testing can be done in five steps.

1. State the problem or question in the form of a Null Hypothesis (H_0).
2. Select a Level of Significance (α).
3. Collect a Sample of data for analysis.
4. Perform a statistical analysis on the Sample data.
5. Come to a conclusion about the Null Hypothesis.

<u>Step 1</u>: State the problem or question in the form of a Null Hypothesis (H_0).
The Null Hypothesis is a statement of nothingness – no difference, change, or effect.

| What's happening? | Absolutely nothin' | No difference | No change | No effect |

The following are some examples.

No (Statistically Significant) Difference

- There is no difference between the Mean heights of Population A and Population B.
- There is no difference between the Standard Deviation of the Process and our target for its Standard Deviation.

No (Statistically Significant) Change

- There is no change between the Mean test scores from last year to this year.
- There is no change in the Mean diameter of holes drilled from the historical Mean.

No (Statistically Significant) Effect

The experimental medical treatment has had no effect on Mean cancer survival rates.

Some experts say that one must also state an Alternative Hypothesis (symbol H_A or H_1) at this point; others disagree.

See the articles *Null Hypothesis* and *Alternative Hypothesis*.

Step 2: **Select a Level of Significance (α).** (covered earlier in this article)

Step 3: **Collect a Sample of data for analysis.**

Not just for Hypothesis Testing, but in general, this must be done <u>after</u> the selection of Alpha. This protects the integrity of the test. If we take a peek at the data first, that might influence our selection of a value for Alpha.

Step 4: **Perform a statistical analysis on the Sample data.**

This can be any kind of Inferential Statistical test.

The test will produce values for the following:

- the Test Statistic, e.g., t
- the Critical Value of the Test Statistic. e.g., t-critical
- p, which is the calculated actual Probability of an Alpha Error

See the articles *Test Statistic, Critical Value*, and *p, p-Value*.

Example output: Hypothesis Test: 2-Sample t-test

t-Stat	3.232		T-Critical Two-tail: 2.145
$P(T<=t)$ two-tail	0.006		Reject Null Hypothesis because $p < 0.05$ (Means are Different)

Step 5: **Come to one of two conclusions about the Null Hypothesis**.

- <u>Either</u> **Reject the Null Hypothesis** if
  Test Statistic $\geq$ Critical Value, which is statistically identical to . . .
  $p \leq \alpha$ (actual Probability is less than or equal to the maximum we said we would tolerate)
- <u>otherwise</u> **Fail to Reject the Null Hypothesis**

These two verbal conclusions can be very confusing, because they involve a double negative and a triple negative, respectively. So, there are articles on each of the two in this book.

A statistician responds to a marriage proposal:

In addition to the articles referenced above, it may be best to read – in fact, start with – the articles *Hypothesis Testing – Part 1: Overview* and *Hypothesis Testing – Part 2: How To.*

 | **5. Inferential Statistics is involved in such analyses as ANOM, ANOVA, Chi-Square Tests, F-tests, Regression, t-tests, and z-tests.**

They all use Sample data and make inferences from them. There are articles on each of these.

In addition to the articles listed above (and in addition to the other articles suggested, in turn, by those articles), it would be a good idea to read the article, *Alpha, p, Critical Value, and Test Statistic – How they Work Together.* It gives a comprehensive explanation of the interactions among these four important numbers in Inferential Statistics, and it should help deepen your understanding.

It explains in detail the concepts and interactions summarized in the following table:

	Alpha	p	Critical Value of Test Statistic	Test Statistic value
What is it?	a Cumulative Probability		a value of the Test Statistic	
How is it pictured?	an $\underline{area}$ under the curve of the Distribution of the Test Statistic		a $\underline{point}$ on the horizontal axis of the Distribution of the Test Statistic	
Boundary	Critical Value marks its boundary	Test Statistic value marks its boundary	Forms the boundary for Alpha	Forms the boundary for p
How is its value determined?	Selected by the tester	area bounded by the Test Statistic value	boundary of the Alpha area	calculated from Sample Data
Compared with	p	Alpha	Test Statistic Value	Critical Value of Test Statistic
Statistically Significant/ Reject the Null Hypothesis if	$p \leq \alpha$		Test Statistic $\geq$ Critical Value e.g., $z \geq z$-critical	

MARGIN OF ERROR

 1. The Margin of Error (MOE) is one-half the width of a 2-sided Confidence Interval.

   A Confidence Interval is a range of values which comprise an Interval Estimate. It can be used in Inferential Statistics instead of the Hypothesis Testing of a Point Estimate. *(See the articles Confidence Intervals – Part 1 and Part 2.)*

   A Confidence Interval is usually described by the value of a Statistic (e.g., the Mean, or Standard Deviation) plus or minus the Margin of Error. For example,

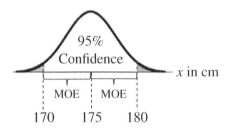

   "MOE" and "E" are used as symbols for Margin of Error.

 2. The "error" is not a mistake.

   It simply the "Sampling Error" – the reduction in accuracy to be expected when one makes an <u>estimate</u> based on a portion (a Sample) of the data in the Population or Process.

 3. Three things affect the size of the MOE: s, α, and n.

   The formula for Margin of Error is:

$$\text{MOE} = \frac{s(\textbf{Critical Value})}{\sqrt{n}}$$

   1. <u>Standard Deviation</u>: s is the Sample Standard Deviation. Being in the numerator – with all other things being equal – **a larger value for the Standard Deviation results in a larger MOE.** This makes sense, because

---

Statistics from A to Z: Confusing Concepts Clarified, First Edition. Andrew A. Jawlik.
© 2016 John Wiley & Sons, Inc. Published 2016 by John Wiley & Sons, Inc.

if the data values Process are spread widely (large Standard Deviation), the Distribution is spread out – and the Confidence Interval is spread out with it, as shown below.

**Smaller Standard Deviation
=> Smaller MOE**

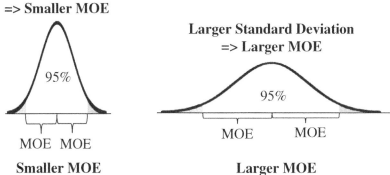

Smaller MOE

**Larger Standard Deviation
=> Larger MOE**

Larger MOE

2. Alpha, α: The Critical Value of a Test Statistic (such z, t, F, or χ^2) is also in the numerator of the formula for MOE. So, larger values for the Critical Value will result in a larger MOE. But, for any given Distribution, the Critical Value is dependent entirely on the value selected for Alpha, the Significance Level. And, as illustrated below, **selecting a smaller value for Alpha results in a larger Critical Value** (farther away from the Mean). **This, in turn, results in a larger MOE.** Alpha is the sum of the two shaded areas under the curves below.

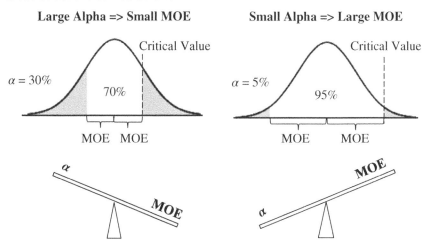

3. Sample Size: The Sample Size, n, is in the denominator of the formula for MOE. So **a larger Sample Size results in a smaller MOE.** There is no statistical tradeoff here. But there may be a practical tradeoff between the increased cost of gathering a larger Sample vs. narrowing the MOE.

Increasing the Sample Size for larger values of n yields diminishing returns in reducing the MOE.

Related Articles in This Book: *Confidence Intervals—Parts 1 and 2; Errors – Types, Uses, and Interrelationships; Standard Deviation; Critical Value; Alpha,α; Test Statistic; Sample Size – Part 2: for Continuous/ Measurement Data*

NONPARAMETRIC

Summary of Keys to Understanding

 1. **Nonparametric (NP) statistical methods have no requirements about the shape of the Distribution(s) from which Samples are collected.**

 2. **Nonparametric methods convert Measurement data into Signs, Ranks, Signed Ranks, or Rank Sums.**

 3. **Nonparametric methods can work with Ordinal data** (e.g., "beginner," "intermediate," "advanced").
NP methods work with Medians instead of Means.

 4. **Nonparametric methods work better in dealing with Outliers, Skewed data, and Small Samples. But, NP methods have less Power.**

 5.

Nonparametric Test	What it does	Parametric Counterpart
Wilcoxon Signed Rank	Compares 1 Median to a specified value	z-test, 1-Sample t-test
	Compares 2 Dependent (Paired) Medians	Paired (Dependent) Samples t-test
Mann–Whitney	Compares 2 Independent Medians	2 (Independent) Samples t-test
Kruskal–Wallis	Compares 3 or more Medians, 1 Variable	1-way ANOVA
Friedman	Compares 3 or more Medians, 2 Variables	2-way ANOVA
Chi-Square Test of Independence	Tests 2 Categorical Variables for Independence (lack of Association)	none

Explanation

> **1. Nonparametric (NP) statistical methods have no requirements about the shape of the Distribution(s) from which Samples are collected.**

Many of the most common statistical analyses have fairly stringent "Assumptions." These are requirements that must be met if the analysis is to be valid. The most common Assumption is that the Population or Process data must have Parameters which approximated those of Normal Distributions.

Parametric Assumption: Normal Distribution

Parameters are statistical properties of a Population or Process. (Corresponding properties of Samples are called Statistics.) The key Parameters which define a Parametric (approximately Normal) Distribution are:

- Mean = Mode = Median
- Skewness = 0 (the shape is Symmetrical)

Nonparametric methods can work with these:

Parametric Assumption: Equal Variance

Parametric methods which use two or more Samples from two different Populations or Processes usually assume roughly equal Variance. Nonparametric methods don't.

Nonparametric methods can work with this:

Nonparametric methods are often called "distribution-free," because they are free of any assumptions about the source Distribution(s).

2. Nonparametric methods convert Measurement data into Signs, Ranks, Signed Ranks, or Rank Sums.

One reason that **NP methods** can use data from any Distribution is that they **usually don't work directly with the data.** The Sample data are converted to Signs and/or Ranks and the numerical values of the data are lost before any calculations are done. So, it makes no difference what type of Distribution the source data have.

Signs

We'll be comparing Sample data to a value we specify. It could be a target, a historical value, an industry standard, etc. Let's say that the historical Median time to complete an operation in an industrial process has been 30 seconds. We collect a Sample of 10 time measurements: 28, 31, 30, 33, 32, 28, 30, 31, 27, 32

If a time is less than 30 seconds, we give it a negative sign. If it is 30 seconds, we give it a zero. If it is greater than 30 seconds, we give it a plus sign.

Specified Value = 30										
Sample Data	28	31	30	33	32	28	30	31	27	32
Sign	−	+	0	+	+	−	0	+	−	+

Count of +'s: 5
Count of −'s: 3

We could use the Counts of these signs – instead of the original data – in a Nonparametric method called the Sign Test.

Ranks

Let's take that same Sample of data, and order it from low to high. Next, assign a Rank from low to high. For ties, split the difference between the values tied. For example, there are two 28's. These occupy two Ranks after 1 (a 2 and a 3), so we give them both a 2.5. The next Rank would be a 4, but there's another tie, so we mark the next two as 4.5's.

Sample Data	27	28	28	30	30	31	31	32	32	33
Rank	1	2.5	2.5	4.5	4.5	6.5	6.5	8.5	8.5	10

Signed Ranks

Signed Ranks, as you might guess, combine the concepts of Signs and Ranks.

But there is a change in how Signs are assigned, and one step uses absolute values, so we'll use a different example with some negative numbers.

Let's say we are doing an analysis of the effect of a training program on employee productivity. If we were doing a Parametric test we'd use the Paired t-test (aka Dependent Samples t-test.) We count the number of transactions that they process in an hour. For each employee, we subtract their Before number from their After number. The data we are capturing is the difference.

Instead of plus and minus signs, we'll use +1 and 0. We compare the data values to a specified value, as we did in our example of the historical Median of 30. Each Sample data value is their After production number minus their Before number.

We'll be testing the Null Hypothesis that there is zero difference, so the specified value is zero.

Step 1: Sign: For each data value, assign a Sign:
 – if it's greater than the specified value (0 in this example), then the Sign $= +1$
 – if it's less than or equal to the specified value: Sign $= 0$

Step 2: Absolute Value

Step 3: Rank the Absolute Values to produce the Absolute Ranks

Step 4: Signed Rank: Multiply the Sign times the Absolute Ranks

Sample data	−6	−2	0	+4	+7	+8	+11	+12	+14	+16
1. Sign	0	0	0	1	1	1	1	1	1	1
2. Absolute Value	6	2	0	4	7	8	11	12	14	16
3. Absolute Rank	4	2	1	3	5	6	7	8	9	10
4. Signed Rank	0	0	0	3	5	6	7	8	9	10

Signed Rank tests are the NP counterpart to the Dependent Samples (aka **Paired Samples**) t-**test.**

Rank Sum tests are the NP counterpart the Independent Samples (aka **2-Samples**) t-**test.**

Rank Sums

We are comparing Samples taken from two Independent Populations or Processes. That is, the data values of one Population are not influenced by the data values of another.

Step 1: Group: Put all the data from both Samples into a single group (but keep track of which ones came from which group).

Step 2: Rank: Rank the values in the combined group.

Step 3: Rank Sum: Total the Ranks for each Sample

1. Data, Sample 1	12	7	15	13	13					
1. Data, Sample 2						11	16	8	6	12
2. Ranks	5.5	2	9	7.5	7.5	4	10	3	1	5.5
3. Rank Sum	31.5					26.5				

There are many Nonparametric methods, and their calculations are usually done with statistical software, so we won't cover that detail here.

 3. Nonparametric methods can also work with Ordinal data (e.g., "beginner," "intermediate," "advanced").
Nonparametric methods work with Medians instead of Means.

Ordinal data are non-numerical; they consist of names. The names imply an ordering (hence "Ordinal"), but there are no corresponding numerical values upon which any calculations can be performed. Even if the names include numbers (e.g., movie ratings of one to four stars) those numbers cannot be used in calculations (a 4-star movie is not defined as 1.33 times as good as a 3-star movie).

The order in Ordinal data is similar to the Ranks in Nonparametric (NP) statistics. So **NP tests which use Ranks are well-suited for Ordinal data.**

Means are normally used as the measure of Central Tendency (aka Central Location) in Parametric tests. In converting the data to Signs, Ranks, etc., we lose the ability to calculate the Mean. **The Median is another measure of Central Tendency. It is well-suited for NP tests, because it uses the number(s) that Rank in the middle.** To determine the Median, arrange the data values in order (low to high or high to low). For an odd number of data values, the Median will be the middle value. For an even number, it will be the average of the two middle values.

The Median has advantages over the Mean:

- **It is less influenced by Outliers.**
- **It is less influenced by the Skew in Skewed data.**

 4. Nonparametric methods work better in dealing with Outliers, Skewed data, and Small Samples. But, NP methods have less Power.

Outliers and Skewed Data

Since **Nonparametric (NP) methods** use the Median instead of the Mean, they **are better than Parametric methods in minimizing the influence of Outliers and Skewed data.**

Also, in certain situations, the Median is a more useful measure of Central Tendency than the Mean. Here's an example: A couple is looking for a house in a community in which most houses are in the price range they can afford, say $400,000. They look at a report on recent home sales which shows Mean prices. One community shows a Mean price of about $670,000, so they immediately exclude it from consideration. But the underlying data show that five houses were sold for around $400,000, and one outlier was sold for $2,000,000. The Median for the same numbers would be around $400,000.

Skewed data:

Similarly, for Skewed data, the Mean can be a less meaningful measure in a practical sense than the Median. Of course, with Skewed data, the Assumption of Normality would not be met, so an NP method would need to be used, in any event.

Small Sample Sizes

For small Sample Sizes, it is not possible to accurately determine whether the Distribution is Normal. So, a small Sample Size is a reason to use an NP method.

Nonparametric methods can be used in any situation where a Parametric method is used. The opposite is, of course, not true. So why wouldn't you use NP methods all the time?

NP methods have less Power. This means that:

- **NP methods have a higher Probability of a Beta (False Negative) Error.**
- **NP methods have less ability to detect small differences, changes, or effects.**

Beta, β, is the Probability of a Beta (False Negative) Error. Power = $1 - \beta$. Power is the Probability of avoiding a Beta Error. Lower Power means that NP methods have a higher Probability of Beta (False Negative) Error than Parametric methods. Also, lower Power means a larger minimum Effect Size. (See the article *Power*.)

Why is this the case? **When Nonparametric methods convert from Measurement data to Signs and Ranks, information – and thus precision – is lost.** The 3rd and 4th ranked Measurements in the data may be 25 and 26, while the 9th and 10th ranked numbers may be 35 and 45. Ranks and signs would lose the information that the difference between the 9th and 10th ranked numbers is 10 times as much as that between the 3rd and 4th. **This reduction in precision is what causes the Power of the test to be reduced in NP methods.**

 5. There are many Nonparametric tests. Here are some of the most commonly used:

Nonparametric Test	What it does	Parametric Counterpart
Wilcoxon Signed Rank	Compares 1 Median to a specified value	z-test, 1-Sample t-test
	Compares 2 Dependent (Paired) Medians	Paired (Dependent) Samples t-test
Mann– Whitney	Compares 2 Independent Medians	2 (Independent) Samples t-test
Kruskal– Wallis	Compares 3 or more Medians, 1 Variable	1-way ANOVA
Friedman	Compares 3 or more Medians, 2 Variables	2-way ANOVA
Chi-Square Test of Independence	Tests 2 Categorical Variables for Independence (lack of Association)	none

Related Articles in This Book: *Normal Distribution*; *Skew, Skewness*; *Power*; *z*; *t-tests-Part 1 and 2*; *ANOVA – Parts 3 and 4*; *Chi-Square Test for Independence*

NORMAL DISTRIBUTION

Summary of Keys to Understanding

 1. Normal Distributions
- **are bell-shaped, left–right symmetrical**
- **have tails which get closer to the horizontal axis, but never touch it**
- **have one Mode and the Mean = Mode = Median**

 2. It takes two Parameters to specify an individual Normal Distribution – the Mean, μ, and the Standard Deviation, σ. The Standard Normal Distribution (whose Test Statistic is z) **has $\mu = 0$ and $\sigma = 1$.**

 3. Empirical Rule: Cumulative Probabilities bounded by Standard Deviations are the same for all Normal Distributions – roughly 68%, 95%, and 99.7% for 1, 2, and 3 Standard Deviations, respectively.

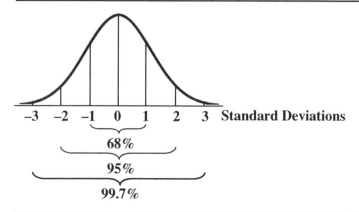

 4. Central Limit Theorem: No matter the Shape of the Distribution of the underlying data, if we take multiple Samples (of the same Size, n) and compute the Means or Proportions for each Sample, **the resulting Distribution of Sample Means or Proportions will be approximately Normal.**

 5. Normality is an Assumption for a number of statistical tests.

Explanation

It may be helpful to first read the article *Distributions – Part 1: What They Are*.

> **1. Normal Distributions**
> - **are Continuous**
> - **are bell-shaped, left–right symmetrical**
> - **have tails which get closer to the horizontal axis, but never touch it**
> - **have one Mode and the Mean = Mode = Median**

Normal (aka Gaussian) Distributions are Continuous and are graphed as smooth curves, like the Exponential, *F*, and Chi-Square Distributions – and unlike Discrete Distributions, such as the Binomial or Poisson.

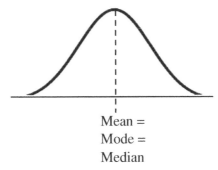

Mean =
Mode =
Median

They are the familiar "bell-shaped curve" with most of their mass centered near the Mean.

Their tails extend asymptotically to infinity to the left and the right. That is, they get ever closer to the horizontal axis without touching it.

There is only one hump designating the Mode. And that same point marks the Mean and the Median – which is required in order to have such a bell shape.

Normal Distributions are by far the most common type of Distribution we run into in statistics and in our daily lives. They are common in natural and human processes that are influenced by many small and unrelated random effects. Some examples:

– height or weight of individuals of the same gender

– test scores

– blood pressure

– Residuals in a Regression Model with a good fit

– variations in a manufacturing process which is under control

Most people are of average height or close to it (the center of the Distribution). The farther one gets from the center out toward the tails of the Distribution – shorter or taller – the smaller the number of people at that height.

> **2. It takes two Parameters to specify an individual Normal Distribution – the Mean, μ, and the Standard Deviation, σ. The Standard Normal Distribution** (whose Test Statistic is z) **has $\mu = 0$ and $\sigma = 1$.**

The Mean, μ, tells us where to position the center of the Normal Distribution – left or right along the Horizontal axis. The Standard Deviation tells us whether the shape of the Normal Distribution will be tall and narrow or wide and short.

small σ large σ

As we'll see next, in the Empirical Rule, it is very useful to select one member of family of Normal Distributions and calculate Probabilities for it. Statisticians made it simple this time, they selected the Normal Distribution with its **Mean = Mode = Median = 0 and the simplest possible Standard Deviation, $\sigma = 1$. This defines the <u>Standard Normal Distribution</u>.**

A Test Statistic is one which has an associated Probability Distribution. z is the Test Statistic associated with the Standard Normal Distribution. (See the article z).

> **3. Empirical Rule: Cumulative Probabilities bounded by Standard Deviations are the same for all Normal Distributions – roughly 68%, 95%, and 99.7% for 1, 2, and 3 Standard Deviations, respectively.**

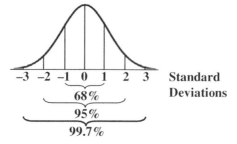

This is also called the "68, 95, 99 Rule" or the "68, 95, 99.7 Rule."

Cumulative Probabilities are Probabilities for ranges of values. They are pictured as areas under the curve. In this case the ranges are between −1 and 1, −2 and 2, and −3 and 3. **These Probabilities are the same for all Normal Distributions.**

The values of the z Test Statistic are plotted along the horizontal axis, and they correspond to the Standard Deviations. **We can calculate the Cumulative Probability for <u>any</u> value of z**, not just integer Standard Deviations. **Likewise, if we know a Cumulative Probability** (as when we select Alpha = 5% in z-test) **we can calculate the value of z** (1.645, for a 1-tailed test).

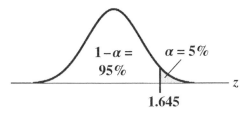

The z transformation formula below can be used to transform any Normal Distribution to the Standard Normal Distribution, aka the z Distribution.

$$z = (x - \mu)/\sigma$$

With a little algebra, we get:

$$x = z\sigma + \mu$$

We know the Cumulative Probability for any value of z. This formula tells us which value of x in another Normal Distribution has that same Cumulative Probability.

---

 4. Central Limit Theorem: No matter the Shape of the Distribution of the underlying data,
    if we take multiple Samples (of the same Size, n) and compute the Means (or Proportions) for each Sample,
    the resulting Distribution of Sample Means (or Proportions) will be approximately Normal.

The Central Limit Theorem (CLT) is a powerful concept which helps us perform statistical analyses on any Population or Process, no matter the underlying shape of its data. We just take Samples of data (with replacement) and calculate the Means for those Samples. **The**

resulting Distribution of these Means of Samples will approximate a Normal Distribution. And then we can use known facts about Normal Distributions – like their Cumulative Probabilities – in statistical analyses.

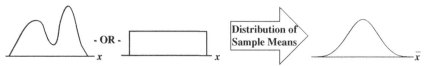

There is something intuitive about the CLT. The Mean of a Sample taken from any Distribution is very unlikely to be at the far left or far right of the range of the Distribution. Means by their very nature tend to average-out extremes, so their Probability would be highest in the center and lowest at the extremes.

Note that x, representing the data, is the horizontal axis Variable for the two strange Distributions left of the arrow. However, **for the Distribution of the Sample Means, the horizontal axis Variable is $\bar{x}$.** The points on the horizontal axis represent values of Means of the Samples taken. (Such a Distribution of a Statistic of a Sample is called a Sampling Distribution; see the article by that name.)

n is the number of individual data points in each Sample; it is not the number of Samples taken. **As n grows larger, the Distribution of the Means of the Samples approaches a Normal Distribution more closely.** For $n \geq 30$, the approximation will be very close; some say that there is no reason to go any higher for the Sample Size. In fact, for use with Control Charts, a Sample Size as low as four can be used.

 5. Normality is an Assumption for a number of statistical tests.

For a number of statistical tests – for example, t-test, z-test, and ANOVA – assume that the data are "roughly Normal." Different tests have different levels of sensitivity to how non-Normal the data can be.

The Anderson–Darling test is often used to determine Normality.

Related Articles in This Book: *Distributions – Part 1: What They Are*; *Distributions – Part 2: What They Do*; *Distributions – Part 3: Which to Use When*; *z*; *Sampling Distribution*; *Control Charts*

NULL HYPOTHESIS

Summary of Keys to Understanding

 1. Stating a **Null Hypothesis and an Alternative Hypothesis is the first step in our five-step method for Hypothesis Testing.**

 2. The Null Hypothesis (symbol H_0) is the hypothesis of <u>nothingness or absence.</u> **In words, the Null Hypothesis is <u>stated in the negative:</u>**

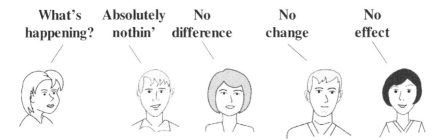

| What's happening? | Absolutely nothin' | No difference | No change | No effect |

 3. It is probably less confusing to state the Null Hypothesis in a formula. It must include an equivalence in the comparison operator, using one of these: "=", "≥", or "≤".

 4. If "=" is not to be used in the Null Hypothesis, start with the Alternative Hypothesis to determine whether to use "≥", or "≤" in the Null Hypothesis.

Comparison Operator		Tails of the Test
H_0	H_A	
$=$	$\neq$	2-tailed
$\leq$	$>$	Right-tailed
$\geq$	$<$	Left-tailed

 5. The last step in Hypothesis Testing is to either
  – "Reject the Null Hypothesis" if $p \leq \alpha$, or
  – "Fail to Reject (Accept) **the Null Hypothesis" if $p > \alpha$.**

Explanation

 | **1.** Stating a **Null Hypothesis and an Alternative Hypothesis is the first step in our five-step method for Hypothesis Testing.**

Hypothesis Testing is one of two common methods for Inferential Statistics. Confidence Intervals is the other. In Inferential Statistics, we estimate a statistical property (e.g., the Mean or Standard Deviation) of a Population or Process by taking a Sample of data and calculating the property in the Sample.

In the article, *"Hypothesis Testing – Part 2: How To "* we describe a 5-step method of Hypothesis Testing:

1. State the problem or question in the form of a Null Hypothesis and Alternative Hypothesis.
2. Select a Level of Significance (α).
3. Collect a Sample of data for analysis.
4. Perform a statistical analysis on the Sample data.
5. Come to a conclusion about the Null Hypothesis (Reject or Fail to Reject).

 | **2. The Null Hypothesis** (symbol H_0) is the hypothesis of **nothingness or absence. In words, the Null Hypothesis is stated in the negative:**

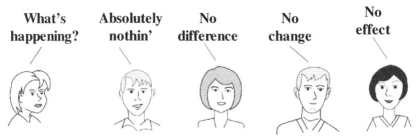

A hypothesis is a statement, opinion, or conjecture about a statistical property of a Population or Process. At the time we state a hypothesis, we don't know if it's true or false. Subsequent steps in the Hypothesis Testing method determine that.

We normally think in terms of things that exist, but the hypothesis in Hypothesis Testing is stated in terms of nonexistence or absence. For example, instead of asking whether the Sample data are Normally distributed, we state a Null Hypothesis like this: "There is no difference between the distribution of this data and a Normal distribution." This is one thing that makes Hypothesis Testing confusing for many people.

Many common Null Hypotheses state that there is <u>no</u> **Statistically Significant**

- **Difference,**
- **Change, or**
- **Effect**

The following are some examples.
No Difference

– There is no difference between the Mean heights of Population A and Population B.
– There is no difference between the Standard Deviation of the Process and our target for its Standard Deviation.

No Change

– There is no change between the Mean test scores from last year to this year.
– There is no change in the Mean diameter of holes drilled from the historical Mean.

No Effect

The experimental medical treatment has had no effect on Mean cancer survival rates.

 3. It is probably less confusing to state the Null Hypothesis in a formula. It must include an equivalence in the comparison operator, using one of these: "=", "≥", or "≤".

In a formula, we don't have to use the confusing language of nonexistence. If we want to say that there is no Statistically Significant difference between the Means of Population A and Population B, we don't have to write it as $\mu_A - \mu_B = 0$. We can just write:

$$H_0: \mu_A = \mu_B$$

That Null Hypothesis would be tested with a 2-sided (2-tailed) test, with $\alpha/2$ under each tail. See the article *Alpha, α,*

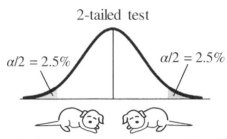

2-tailed test

$\alpha/2 = 2.5\%$ $\alpha/2 = 2.5\%$

But, we may not be interested in <u>whether</u> there is a difference, so much as if there is a difference <u>in a particular direction</u>. For example, if we make light bulbs with an advertised minimum life of 1300 hours, we don't want to know whether the actual Mean (as estimated from Sample data) within a statistically insignificant distance from 1300 hours. We want to know if the actual Mean is 1300 <u>or more</u>. Our Null Hypothesis would then be:

$$H_0: \mu \leq 1300 \text{ hours}$$

And if there was a Statistically Significant conclusion from the test, we would Reject this Null Hypothesis and conclude that the $\mu > 1300$ hours.

But this is kind of confusing: the $\leq$ in H_0 points in the opposite direction from the $>$ in the conclusion. That's why we use the Alternative Hypothesis, H_A, instead of the Null Hypothesis, to determine the direction of the tail in 1-tailed analyses. One-tailed analyses can be either left- or right-tailed. A shaded area representing the full value of α is under either the left or right tail.

> **4. If "=" is not to be used in the Null Hypothesis, start with the Alternative Hypothesis, H_A, to determine whether to use "$\geq$", or "$\leq$" in the Null Hypothesis.**
> **H_A also tells us which way the tail points in 1-tailed tests.**

- The Alternative Hypothesis (notation: H_A or H_1) is the opposite of the Null Hypothesis.
- It is the Hypothesis which is true when the Null Hypothesis is false – and vice versa.
- Whereas the Null Hypotheses always has an equivalence in its comparison operator, the Alternative Hypothesis never does. The Null hypothesis has "$<$" or "$>$".

But the biggest benefit to stating an Alternative Hypothesis is:

The Alternative Hypothesis points in the direction of the tail in a 1-tailed test.

That is why – **for a 1-tailed test – it is less confusing to start by stating an Alternative Hypothesis and then take the opposite of that as the statement of the Null Hypothesis**.

There is more on this in the article *Alternative Hypothesis*, but the following table may be helpful in explaining this relationship between H_0, H_A and the tails of the test.

Comparison Operator		Tails of the Test	
H_0	H_A		
$=$	$\neq$	2-tailed	$\alpha/2$ $\alpha/2$
$\leq$	$>$	Right-tailed	$\alpha = 5\%$
$\geq$	$<$	Left-tailed	$\alpha = 5\%$

> **5. The last step in Hypothesis Testing is to either**
> – **"Reject the Null Hypothesis"** if $p \leq \alpha$, or
> – **"Fail to Reject** (Accept) **the Null Hypothesis"** if $p > \alpha$.

Note that $p \leq \alpha$ is statistically identical to the value of the Test Statistic being $\geq$ the Critical Value (e.g., $t \geq t$-critical).

If the Null Hypothesis is stated as a negative, then to "Reject" it is a double negative, and to "Fail to Reject" it is a triple negative. This gets confusing, even for experienced practitioners.

To help clear up the confusion, you can read the two articles in this book which have as their titles the two alternatives for the last step, listed above.

And, as we note in the article *Hypothesis Testing – Part 1*, some practitioners choose to avoid Hypothesis Testing and use Confidence Intervals instead.

Related Articles in This book: The Null Hypothesis is just one concept in Hypothesis Testing. We've broken things up into bite-sized-chunks, consisting of the following articles.

- *Hypothesis Testing – Part 1: Overview*
- *Hypothesis Testing – Part 2: How To*
- *Null Hypothesis (this article)*
- *Reject the Null Hypothesis*
- *Fail to Reject the Null Hypothesis*
- *Alternative Hypothesis*

Also, ...
Alpha, α; Confidence Intervals – Parts 1 and 2; Inferential Statistics

_p_, _p_-VALUE

Summary of Keys to Understanding

 1. In Inferential Statistical analyses, _p_ is the Probability of an Alpha Error (also known as a "False Positive" or a Type I Error).

 2. From Sample data, a value is calculated for a Test Statistic. This value is plotted on the Probability Distribution of the Test Statistic. **The _p_-value is calculated as the Cumulative Probability of the area under the curve beyond the Test Statistic Value.**

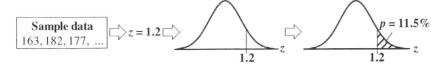

| Sample data |
| 163, 182, 177, ... |

$\Rightarrow z = 1.2 \Rightarrow$ $p = 11.5\%$

 3. In Hypothesis Testing, _p_ is compared with Alpha to determine the conclusion from an Inferential Statistics test. If $p \leq \alpha$, Reject the Null Hypothesis. If $p > \alpha$, Fail to Reject (i.e., Accept) the Null Hypothesis

Areas under the curve (right tail) α: ▭ p: ▨	z-critical	z-critical
	$p \leq \alpha$ $z \geq z$-critical	$p > \alpha$ $z < z$-critical
The observed difference, change, or effect is:	Statistically Significant	not Statistically Significant
Null Hypothesis	Reject	Accept (Fail to Reject)

 4. The smaller the value of _p_, the more accurately the Sample represents the Population or Process.

Statistics from A to Z: Confusing Concepts Clarified, First Edition. Andrew A. Jawlik.
© 2016 John Wiley & Sons, Inc. Published 2016 by John Wiley & Sons, Inc.

Explanation

> **1. In Inferential Statistical analyses, *p* is the Probability of an Alpha Error** (also known as a "False Positive" or a Type I Error).

In Inferential Statistics, we use data from a Sample to estimate a property (e.g., the Mean) of the Population or Process from which the Sample was drawn. Being an estimate, there is a chance for error.

In Inferential Statistical tests – such as the *z*-test, the *t*-tests, the *F*-tests, Chi-square tests, and ANOVA – **the conclusion we make from the test depends on how likely it is that we have avoided an Alpha Error.**

An Alpha Error is the error of observing something – for example, a change in a Process, **a difference** between two treatments, or **an effect** (positive or negative) of a new drug – **when there is nothing**. It may be more memorable to think of an Alpha Error as a False Positive, like this one:

I saw a unicorn.

Alpha Error
(False Positive)

In Hypothesis Testing (one of the two main methods of Inferential Statistics), **it is the error of rejecting the Null Hypothesis** (and concluding that there <u>is</u> a difference, change, or effect) **when the Null Hypothesis** (which says there is <u>not</u>) **is true**. *(See the article "Alpha and Beta Errors.")*

> **2.** From Sample data, a value is calculated for a Test Statistic. This value is plotted on the Probability Distribution of the Test Statistic. **The *p*-value is calculated as the Cumulative Probability of the area under the curve beyond the Test Statistic Value.**

A Statistic is a numerical property of a Sample, e.g., the Mean or Standard Deviation. A Test Statistic is a Statistic that has an associated

Probability Distribution. The most common are z, t, F and Chi-Square. Below is the formula for z, which is used in analyses of Means:

$$z = (\mu - \bar{x})/s$$

where $\bar{x}$ is the Sample Mean and s is the Standard Deviation and μ is a specified value of the Population Mean. It could be an estimate, a historical value, or a target, for instance.

Here's how the value of *p* (the "*p*-value") is determined:

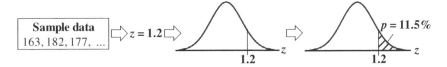

- **The Sample Data are used to calculate a value for the Test Statistic** (1.2 in this example).
- **This Test Statistic value is plotted on the graph of the Probability Distribution of the Test Statistic.**

The height of the curve above each value on the horizontal axis is the Probability of that value occurring. **The Cumulative Probability of a range of values occurring is the area under the curve above those values.**

- ***p* is calculated** (from tables or software) **as the Cumulative Probability of the range of values from the Test Statistic value outward** (from the Mean, which in this case is to the right, extending to infinity.)

 3. In Hypothesis Testing, *p* is compared with Alpha to determine the conclusion from an Inferential Statistics test. If $p \leq \alpha$, Reject the Null Hypothesis. If $p > \alpha$, Fail to Reject (i.e., Accept) the Null Hypothesis

(If you're not familiar with the concept of Alpha, it may be a good idea to read the article *Alpha, α* before proceeding.) We select a value for Alpha, the Level of Significance, before we collect a Sample of data. If we want a 95% Level of Confidence of avoiding an Alpha Error, then we would select $100\% - 95\% = 5\%$ (0.05) for Alpha.

Alpha is called the Level of Significance, because **if the Probability (*p*) of an Alpha Error is less than or equal to our selected level of Alpha, we can call the results of our test "Statistically Significant."**

Areas under the curve (right tail) α: ☐ p: ▨	z-critical $\underset{z}{|}$	z-critical $\underset{z}{|}$
	$p \leq \alpha$ $z \geq z$-critical	$p > \alpha$ $z < z$-critical
The observed difference, change, or effect is:	Statistically Significant	not Statistically Significant
Null Hypothesis	Reject	Accept (Fail to Reject)

Since *p* is defined as the Probability of an Alpha Error, $p \leq \alpha$ means that any difference, change, or effect observed in our Sample Data is Statistically Significant, and we Reject the Null Hypothesis (which states that there is no difference, change, or effect).

$p \leq \alpha$ is shown in the diagram in the middle column of the table above, which is a closeup of the right tail of the curve of the Test Statistic (z, in this case) Distribution. Our example, in which $p = 11.5\%$ and $\alpha = 5\%$, is illustrated by the right column.

> **4. The smaller the value of *p*, the more accurately the Sample represents the Population or Process.**

Since *p* is the Probability of an error, it **has an inverse relationship to the validity of the Statistical Inference.** If *p* is high, then the Probability for error is high, and it is unlikely that the estimate from the Sample is an accurate portrayal of the Population or Process. Conversely, if *p* is low, then the Probability for error is low, and it is likely that the estimate is accurate.

Instead of selecting a value for Alpha and then comparing *p* to that value, some experimenters prefer to use *p* alone. Various levels of strength of evidence are defined, as shown below. Different experimenters and different disciplines may use somewhat different adjectives and different clip levels. Here is one scheme:

$p > 0.1$ is very weak or no evidence against the Null Hypothesis.

$0.05 < p < 0.1$ is slight or weak evidence against the Null Hypothesis.

$0.01 < p < 0.05$ is moderate evidence against the Null Hypothesis.

$0.001 < p < 0.01$ is strong evidence against the Null Hypothesis.

$p < 0.001$ is very strong or overwhelming evidence against the Null Hypothesis.

And finally, to put it all in context, below is a compare-and-contrast table showing how *p* fits in with and interacts with three other key elements of Inferential Statistics. This diagram is explained in detail in the article *Alpha, p-Value, Test Statistic, and Critical Value – How They Work Together*

	Alpha	***p***		**Critical Value of Test Statistic**	**Test Statistic value**
What is it?	a Cumulative Probability			a value of the Test Statistic	
How is it pictured?	an <u>area</u> under the curve of the Distribution of the Test Statistic			a <u>point</u> on the horizontal axis of the Distribution of the Test Statistic	
Boundary	Critical Value marks its boundary	Test Statistic value marks its boundary		Forms the boundary for Alpha	Forms the boundary for *p*
How is its value determined?	Selected by the tester	area bounded by the Test Statistic value		boundary of the Alpha area	calculated from Sample Data
Compared with	*p*	Alpha		Test Statistic Value	Critical Value of Test Statistic
Statistically Significant/ Reject the Null Hypothesis if	$p \leq \alpha$			Test Statistic ≥ Critical Value e.g., $z \geq z$-critical	

Related Articles in This Book: *Alpha, p-Value, Test Statistic, and Critical Value – How They Work Together*; *Alpha, α*; *Alpha and Beta Errors*; *Distributions – Part 1: What They Are*; *Inferential Statistics*; *Test Statistic*; *Hypothesis Testing – Part 1: Overview*; *Null Hypothesis*; *Reject the Null Hypothesis*; *Fail to Reject the Null Hypothesis*; *p, t, and F: ">" or "<"?*

p, t, AND *F*: ">" OR "<"?

We run into *p* and Test Statistics – such as *t*, *F*, *z*, and χ^2 – in a number of statistical tests, such as *t*-tests, *F*-tests, and ANOVA. After performing one of these tests, we come to a conclusion based on **whether $p \leq$ or > 0.05** (or other value for Alpha) – **or whether the Test Statistic is greater than or less than its Critical Value.**

But beginners can sometimes forget which way the "<" or ">" is supposed to point in each case. In this article, we'll clarify this in three different ways – first, via three rules, second, by providing a Statistical Explanation, and third, with a gimmicky memory cue.

1. Three Rules

Rule #1: *t*, *F*, *z*, and χ^2 – all point in the same direction.
  They are all Test Statistics and behave similarly. **That is, $t \geq t$-critical, $F \geq F$-critical, $z \geq z$-critical, and $\chi^2 \geq \chi^2$-critical all imply the same conclusion: Reject H_0, the Null Hypothesis.** That is, we conclude that **there is a Statistically Significant difference, change, or effect.**

Rule #2: *p* points in the opposite direction.
  p is not a Test Statistic. **$p \leq \alpha$ means the same as $t \geq t$-critical, $F \geq F$-critical, etc.**

  So, a small value for *p* corresponds to a large value for *t*, *F*, *z*, and χ^2, and vice-versa

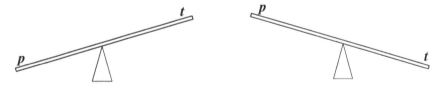

  Alpha, α, is the Level of Significance (1- the Level of Confidence) which we select. Most commonly $\alpha = 0.05$ (5%) is selected.
  Rules #1 and #2 are illustrated in the following output from an ANOVA analysis.

ANOVA Table	$\alpha = 0.05$					
Source of Variation	**SS**	**df**	**MS**	**F**	**p-value**	**F-crit**
Sample	103	1	103	5.9	0.031	4.7
Columns	254	2	127	7.4	0.008	3.9
Interaction	20	2	10	0.6	0.574	3.9

The three rows for the Variables, "Sample," "Column," and "Interaction" each have p-values as well as values for the Test Statistic, F, and its Critical Value ("F-crit").

- Sample: p at 0.031 $< \alpha$ at 0.05 and F at 5.9 $> F$-crit at 4.7
- Columns: p at 0.008 $< \alpha$ at 0.05 and F at 7.4 $> F$-crit at 3.9
- Interaction: p at 0.574 $> \alpha$ at 0.05 and F at 0.6 $< F$-crit at 3.9

Rule #3: **Reject H_0 if $p \leq \alpha$; Accept** (Fail to Reject) **H_0 if $p > \alpha$.**

(You can forget the Test Statistic, and use only p. This is because of Rule 2, above.)

2. Statistical Explanation

These graphs are close-ups of the right tail of a t Distribution for a 1-tailed test, but the principle applies for all Test Statistics. p and Alpha are Cumulative Probabilities represented by areas under the curve. Alpha is the shaded area under the curve. p is the hatched area.

Areas under the curve (right tail)

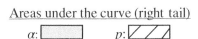

t-critical marks the boundary for α, and t marks the boundary for p.

t-critical t t t-critical

$t \geq t$-critical, so $p \leq \alpha$ $t < t$-critical so $p > \alpha$

Left diagram above: $t \geq t$**-critical so $p \leq \alpha$** (these two comparisons are statistically identical)

The value of the Test Statistic, e.g., t, is calculated from the Sample data. It is plotted on the horizontal axis below the curve of the Test Statistic Distribution. The value of p is then calculated as the area under the curve beginning at t and extending outward below the tail of the curve.

The larger the value for t (the farther to the right it is on the horizontal axis), the smaller the area representing p. So, if t is greater than or equal to t-critical, then the hatched area representing p is less than or equal to the shaded area representing Alpha (which is bounded by t-critical). So, $t \geq t$**-critical means that $p \leq \alpha$.**

So, we must conclude that there **is a Statistically Significant difference, change, or effect.** In Hypothesis Testing, we **Reject the Null Hypothesis** of no difference.

Right Diagram above: *t* < *t*-**critical so** *p* > *α* (these two comparisons are statistically identical)

If the calculated value of *t* is < *t*-**critical**, then *t* is to the left of *t*-critical on the horizontal axis. This means that the area under the curve bounded by *t* is greater than the area bounded by *t*-critical. So *p* > *α*.

In Hypothesis Testing, we **Fail to Reject** (i.e., we **Accept) the Null Hypothesis**, which says there is no difference, change, or effect.

Another statistical explanation which may be helpful: *p* **represents error, and we would like** *p* **to be small before we make a claim of Statistically Significant.**

3. Memory Cue

(This is a non-statistical gimmick, but it may be helpful for some – it was for the author.) In this book, we don't focus on confusing things like the "nothingness" of the Null Hypothesis. We focus on something that <u>does</u> exist – like a difference, change, or effect. So, to make things easy, we want a memory cue that tells us when there <u>is something</u>, as opposed to nothing.

We can come to the following conclusions (depending on the test):

If	then
$p \leq \alpha$, $t \geq t$-**critical,** $F \geq F$-**critical, etc.**	there <u>**is**</u> **a difference** between the two Samples, or something **has changed** in the process, or the treatment we are studying <u>**does**</u> **have an effect**. So, we **Reject H$_0$**, the Null Hypothesis.

But how do we remember which way the inequality symbol should go?

Remember back in kindergarten or first grade, when you were learning how to print? The letters of the Alphabet were aligned in three zones – middle, upper, and lower as below.

p is different from *t* or *f*, because *p* **extends into the lower zone**, while *F*, *t*, **and** χ^2 **extend into the upper zone.** (*z* doesn't; it stays in the middle zone. But we can remember that *z* is similar to *t* and it is a Test Statistic.)

If we associate the **lower zone** with **less than**, and the **upper zone** with **greater than**, we have the following memory cue:

Related Articles in This book: *Alpha (α)*; *p, p-Value*; *Null Hypothesis*; *Reject the Null Hypotheses*; *Fail to Reject the Null Hypothesis*; *Test Statistic*; *Critical Values*; *Alpha, p, Critical Value and Test Statistic – How They Work Together*

POISSON DISTRIBUTION

Summary of Keys to Understanding

 1. The Poisson Distribution is a Distribution for **Discrete data**. It consists of the **Probabilities of Counts of Occurrences.**

 2. **Given an average number of Occurrences, the Poisson Distribution is useful in** <u>predicting</u> **the Probability of X (or more than X or fewer than X) Occurrences in a given observation area in** time or space.

 3. There are **different Poisson Distributions for different values of the Mean, μ.**

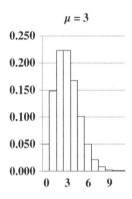
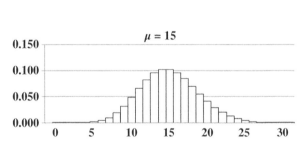

 4. **The Variance (σ^2) of a Poisson Distribution is equal to the Mean (μ) and the Expected Value (λ).**

$$\sigma^2 = \mu = \lambda$$

 5. **The Binomial Distribution converges to the Poisson Distribution as its number of trials, n, approaches infinity.**

Explanation

 1. The Poisson Distribution is a Distribution for **Discrete data.** It consists of the **Probabilities of Counts of Occurrences.**

Discrete data are integers, such as Counts. Counts are non-negative. In contrast to Continuous data, there are no intermediate values between consecutive integer values.

The Poisson Distribution is used for Counts of Occurrences, as opposed to Units.

Occurrences are different from Units. Let's say we are inspecting shirts at the end of the manufacturing line. We may be interested in the number of defective Units – shirts, because any defective shirt is likely to be rejected by our customer. However one defective shirt can contain more than one defect. So, we are also interested in the Count of individual defects – the Occurrences – because that tells us how much of a quality problem we have in our manufacturing process.

Other Discrete data Distributions, the Binomial and the Hypergeometric, are used for Units, not Occurrences.

If the Occurrences are rare, instead of using Counts and the Poisson Distribution, record the time between Occurrences, and use the Exponential Distribution.

For an example of relatively rare Occurrences, let's say that a small town averages 0.4 ambulance calls per day. In trying to determine the Probability of a call during an upcoming understaffed 2-day period, it would be better to think in terms of the average time-to-occurrence.

0.4 Occurrences/24 hours = 0.017 Occurrences per hour = 59 hours between Occurrences. They could use this number and the Exponential Distribution to calculate the Probability of a call during the 48-hour period.

 2. **Given an average number of Occurrences, the Poisson Distribution is useful in <u>predicting</u> the Probability of X (or more than X or fewer than X) Occurrences in a given observation area** in time or space.

For example, let's say we are operating a small call center which averages 3 incoming calls every 10 minutes. We can handle up to 5 calls in 10 minutes. What is the Probability that our capacity will be exceeded?

The Poisson Distribution can tell us the Probability of exactly $X = 6$ in 10 minutes. But, we also need to know the Probabilities of 7, 8, and so on. The Poisson Distribution can tell us those also.

But, since there is theoretically no limit on the number of calls, our approach is to use the Poisson Distribution to get the Probabilities of $X = 0, 1, 2, 3, 4$, and 5. We total these and subtract from 1 (or 100%) to get the Probability of exceeding 5.

$Pr(X = 0)$: 0.050
$Pr(X = 1)$: 0.149
$Pr(X = 2)$: 0.224
$Pr(X = 3)$: 0.224
$Pr(X = 4)$: 0.168
$Pr(X = 5)$ 0.101

The Probability of X being five or fewer is the sum of these six Probabilities, so $Pr(X \leq 5) = 0.916$. So the Probability of exceeding our limit of five calls is $1 - 0.916 = 0.084$, about 8%.

 | **3. There are different Poisson Distributions for different values of the Mean, μ.**

We need only one Parameter, the Mean, to define a Poisson Distribution, and different values of the Mean give us different Distributions:

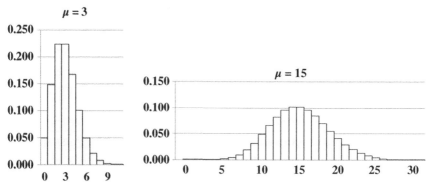

Notice that, as the Mean grows larger:

- the Distribution moves to the right (because the Mean moves to the right)
- The individual Probabilities get smaller (the bars are shorter)
- the Distribution becomes more symmetrical
- the Distribution spreads out (the Variance grows larger)

Alternate terminology and symbols: The Mean of a Poisson Distribution is also known as the Expected Value $E(X)$. It is also sometimes denoted by λ, instead of μ.

---

4. The Variance (σ^2) of a Poisson Distribution is equal to the Mean (μ) and the Expected Value (λ).

$$\sigma^2 = \mu = \lambda$$

That explains why the Distribution spreads out as the Mean gets larger, as shown in the graphs above.

The Population or Process Mean is the average. The Expected Value (λ) is similar to a Sample average. It is based on past data, and it gives us an estimate of what to expect in the future. If our store had 71 customers in a typical 10-hour day, the Mean would be 7.1 customers per hour. And we might expect around 7.1 customers per hour going forward.

---

5. The Binomial Distribution converges to the Poisson Distribution as its number of trials, *n*, approaches infinity.

The Binomial Distribution is another Discrete data Distribution. *n* is its number of trials (a trial can be a coin flip).

One could say that the Poisson Distribution is a limiting case of the Binomial Distribution or that the Binomial is a special case of the Poisson. The Binomial has two Parameters, *n* and *p*, which is the Probability of a trial. If *p* remains fixed and *n* increases, the Binomial converges to the Poisson. For large *n* and small *p*, one Distribution has been used as an approximation for the other.

Related Articles in This book: *Distributions – Parts 1–3*; *Variance; Binomial Distribution*

POWER

It may be helpful to first read the article on Alpha and Beta Errors.

Summary of Keys to Understanding

 1. **The Power of a test** in Inferential Statistics **is its Probability of correctly concluding** that there is a difference, change, or effect, when, in reality, there is one.

 2. **Power is the opposite of Beta.** Beta (β) is the Probability of making a Beta (False Negative) Error. **Power is the Probability of not making a Beta Error. Power = 1 − β**

 3. **Power is affected by three factors – directly by Significance Level (α) and Sample Size (n), and inversely by the Effect Size (ES).**

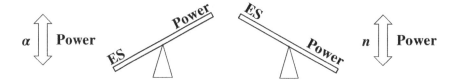

 4. **Power is useful for determining the minimum Sample Size needed to detect an effect of a specified Effect Size.**

 5. **In the social and behavioral sciences, a minimum Power of 80% has been suggested. In medical studies and manufacturing quality studies, higher Powers are needed.**

 6. **Parametric Tests**, e.g., the *F*-test, **have higher Power than Nonparametric Tests**, e.g., Levene's test.

Explanation

It may be helpful to first read the article on Alpha and Beta Errors.

> **1. The** <u>Power of a test</u> in Inferential Statistics **is its** <u>Probability</u> <u>of correctly concluding</u> **that there is no difference, change, or effect when**, in reality, **there is none**.

Put in other words:

The <u>Power of a test</u> in Inferential Statistics **is its** <u>Probability of</u> <u>correctly rejecting</u> **the Null Hypothesis when it is false.**

Power is good. Higher Power gives you a more accurate test.

> **2. Power is the opposite of Beta.** Beta (β) is the Probability of making a Beta (False Negative) Error. **Power is the Probability of <u>not</u> making a Beta Error. Power = 1 − β**

In Inferential Statistical tests – such as the *t*-tests or ANOVA – a property (e.g., the Mean) of the Sample data is used to <u>estimate</u> the corresponding property of the Population or Process from which the Sample was collected.

Since it is an estimate, there are errors involved. These are expressed as Alpha and Beta Errors (False Positive and False Negative, respectively).

	Beta Error (False Negative)
	Smoking doesn't cause cancer.
What it means	**The error of** <u>concluding that there is</u> <u>nothing</u> – no difference, or no change, or no effect – **when, in reality, there is**
Null Hypothesis	**is Accepted (Fail to Reject) when it is false**
Probability of making this Error	β
Probability of <u>not</u> making this Error	**Power = 1 − β**

If you are familiar with the concept of Alpha and Alpha Errors, you will need to be a little careful with the terms associated with Beta and Beta Errors. There is not symmetry in these terms. β is analogous to p, not α:

p is the Probability of an Alpha Error	β is the Probability of a Beta Error
α is the maximum tolerable Probability for an Alpha Error	
$1 - \alpha$ is called the Confidence Level	$1 - \beta$ is called the Power of the test

> **3. Power is affected by three factors – directly by Significance Level (α) and Sample Size (n), and inversely by the Effect Size (ES).**

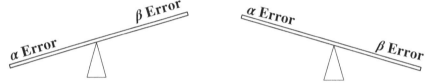

The illustrations above are from the article *Alpha and Beta Errors*. All other things being equal, as the Probability of an Alpha Error decreases, the Probability of a Beta Error increases.

If we require a low Probability of an Alpha Error (at left), we select **a low value for Alpha (α). This causes β to increase. If β increases, then Power** (which is $1 - \beta$) **decreases.**

So, **decreasing Alpha causes Power to decrease** (if Effect Size and Sample Size remain the same). And, **increasing Alpha causes Power to increase.**

α Power

Power and Effect Size, on the other hand, **have an inverse relationship.**

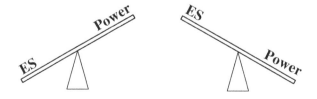

If we want to be able to detect a small difference, change or effect (small Effect Size), **then we'll need a test with more Power than we would need for larger Effect Sizes.** This can be achieved by increasing the Sample Size.

Samples Size (*n*) affects Power directly. All other things being equal, increasing the Sample Size increases the Power of the test. Reducing the Sample Size reduces the Power.

n **Power**

We keep saying "all other things being equal." But what if all other things are not equal? What if we wanted to reduce the Effect Size, reduce the Alpha Error, <u>and</u> increase the Power. **Sample Size is the universal cure.** Keep increasing the Sample Size, and you can counteract both of the other things that go into determining the value of Power.

 | **4. Power is useful for determining the <u>minimum Sample Size</u> needed to detect an effect of a specified Effect Size.**

After an Inferential Statistics test is completed ("*post hoc*"), Power is not very useful for analyzing the results. In fact, there is controversy regarding whether it should be used then at all.

However **before the data are collected** ("a priori") **Power can be used to determine the minimum Sample Size needed to detect a difference, change, or effect of a given Effect Size.**

We said earlier that Power is affected by three things, Alpha, Effect Size, and Sample Size. In other words, **Power is a function of Alpha, Effect Size, and Sample Size.**

$$\text{Power} = f(\text{Alpha, Effect Size, Sample Size})$$

The actual formula is complicated beyond the scope of this book. But if we did some math, we could manipulate it to form a function for Sample Size:

$$\text{Sample Size} = f(\text{Power, Alpha, Effect Size})$$

Minimum Sample Size calculations from these three things can be done with commercial or open source software or via website calculators.

| **5. In the social and behavioral sciences, a minimum Power of 80% has been suggested. In medical studies and manufacturing quality studies, higher Powers are needed.**

In the social and behavioral sciences, it has been said that it is about four times as important to avoid a False Positive as it is to avoid a False Negative. So, if Alpha is selected to be 0.05, then a maximum value for Beta would be 0.20. Power $= 1 - \beta = 0.8 = 80\%$. And the Sample Size would be calculated using these requirements. In practice, a number of studies in these areas focus entirely on avoiding Alpha Errors in their experimental design. The resulting studies then have Power considerably less than 80%.

In medical studies, it is very important to avoid a False Negative (concluding that there is no disease, when in reality there is), so the required Power will be much higher.

In process manufacturing quality studies, the specification limits which define the acceptable range of measured values can be quite narrow. So, the Effect Size is small. Consequently, the Power of the test must be high in order to detect small effects. How high is determined by Power and Sample Size software into which Alpha and Effect Size have been entered as inputs.

 | **6. Parametric Tests**, e.g., the *F*-test, **have higher Power than Nonparametric Tests**, e.g., Levene's test.

Parametric tests are the most commonly used tests. But they can have somewhat stringent requirements (assumptions) – Normal distribution and equal Variance, for example. There are "Nonparametric" counterparts to these tests which do not have those assumptions. The natural question is Why not use Nonparametric tests all the time? The answer is that they have lower Power than Parametric tests.

Related Articles in This Book: *Alpha and Beta Errors*; *Alpha, p-Value, Critical Value, and Test Statistic – How They Work Together*; *Design of Experiments – Part 1*; *Nonparametric*; *Sample Size – Parts 1 and 2*

PROCESS CAPABILITY ANALYSIS (PCA)

Summary of Keys to Understanding

 1. First, use Control Charts and Run Rules to determine whether the Variation in a Process is stable and predictable within statistically calculated Control Limits.

	Tools Used	Focus	Variation Limits
Statistical Process Control (SPC)	Control Charts, Run Rules	eliminate Special Cause Variation	calculated Control Limits
Process Capability Analysis (PCA)	**Capability Indices, Performance Indices**	**reduce Common Cause Variation**	**specified Specification Limits**

 2. If the Process is Under Control, then Process Capability Analysis (PCA) can be used to determine how consistently Variation in the Process is contained within Specification Limits.

$$\text{Process Capability} = \frac{\text{Voice of the Customer}}{\text{Voice of the Process}}$$

 3. C_{pk} is a Capability Index which measures the best that the Process is capable of – in terms of short-term Variation.

$$C_{pk} = \frac{\text{Absolute Value (Mean – Closer Spec Limit)}}{3\hat{\sigma}}$$

 4. For every Capability Index, there is a corresponding Performance Index which measures how well the Process actually performs over the long term.

 5. Capability Indices are used to calculate the Sigma quality level of a Process and to guide efforts to improve the Process.

Explanation

> **1. First, use Control Charts and Run Rules to determine whether the Variation in a Process is stable and predictable within statistically calculated Control Limits.**

- All Processes have Variation. Control Charts and Run Rules (see the articles *Control Charts – Parts 1 and 2*) tell us whether this Variation is stable and predictable, that is, whether the Process is "Under Control."
- **If a Process is <u>not</u> Under Control, we <u>cannot</u> use Process Capability Analysis (PCA).** That is because we really don't know what the actual Process is. Known or unknown Factors (Special Causes of Variation) <u>outside</u> the designed Process are changing what should normally happen within it.
- These **Special Causes of Variation must be identified and eliminated before we can use PCA to begin efforts to reduce Variation <u>within</u> the Process**

Variation from Special Causes

- is identified by Control Charts and Run Rules
  - Variation outside the Control Limits on the Control Charts
  - non-random patterns identified by Run Rules
- is unexpected
- is a "signal" that something is different or has changed (as opposed to the "noise" of random Variation within a Process which is Under Control)
- is outside the Process

Upper and Lower **Control Limits** for Control Charts **are <u>calculated</u>**. Usually, they are plus and minus three Standard Deviations (Sigmas) from a Center Point (usually the Mean). This would include about 99.7 of expected random Variation in Normally distributed data.

	Tools Used	**Focus**	**Variation Limits**
Statistical Process Control (SPC)	Control Charts, Run Rules	eliminate <u>Special Cause</u> Variation	calculated Control Limits
Process Capability Analysis (PCA)	**Capability Indices, Performance Indices**	**reduce <u>Common Cause Variation</u>**	**<u>specified</u> Specification Limits**

| **2. If the Process is Under Control, then Process Capability Analysis (PCA) can be used to determine how consistently Variation in the Process is contained within Specification Limits.**

If there is no Special Cause Variation, then any remaining Variation is from Common Causes. This is the normal, expected random "noise" within any Process – for example, the normal tiny amount of wobbling of a drill in a drilling operation.

Specification Limits specify upper and/or lower bounds on a key measurement in a Process. They can be specified by:

- Customer requirement
- Management decision
- Engineering requirement

For example, let's say we make quarter-pound ground beef patties for sale to restaurants. Our customers tell us that anything less than 0.23 pounds is unacceptable. Our management tells us that anything over 0.27 pounds is unacceptable waste. So, our lower and upper Specification Limits are 0.23 and 0.27 pounds.

In this example, the Lower Specification Limit (LSL) was set by the customer and the Upper Specification Limit (USL) by management. Spec Limits can also be specified by engineering requirements. For example, a part produced by one step in a manufacturing process must measure within specified limits in order to be usable by a machine in a subsequent process step. Spec Limits can be upper or lower, or both.

So, how can we measure how consistently our Process is meeting the Specification Limits? That is what Capability Indices and Performance Indices do.

Use Measurements, not Counts

As we'll see later, PCA Indices can be used to estimate the defects per million opportunities (DPMO) in a Process. But, in capturing the data, be sure to preserve the values of the measurements. **Do not use a measurement to determine a Count of defect or no defect; to do so would lose information and reduce accuracy.**

In our quarter pound burger example, if we measured burgers weighing 0.28 and 0.33 pounds (both outside the Spec Limits), we should not record them as a Count of 2 defects. That would treat these two equally, when in fact, there is a substantial difference between the two. We need to record the actual measurements – 0.28 and 0.33.

| **3. C_{pk} is a Capability Index which measures the best that the Process is capable of – in terms of short-term Variation.**

Conceptually,

$$\textbf{Process Capability} = \frac{\textbf{Voice of the Customer}}{\textbf{Voice of the Process}}$$

The "Voice of the Customer" (or Management or Engineering) is defined by the Specification Limits. It is the difference between the Upper Specification Limit and the Lower Specification Limit, that is, USL – LSL.

The "Voice of the Process" is defined as the width of the Range within which 99.7% of the Process data measurements fall in Normally distributed data. This is three Standard Deviations (3σ) above and 3σ below the Mean, for a total of 6σ. The narrower the Range the better. Conceptually,

$$\text{Process Capability} = \frac{\text{Voice of the Customer}}{\text{Voice of the Process}} = \frac{\text{USL} - \text{LSL}}{6\sigma}$$

Since we're usually analyzing ongoing Processes, potential new data continue to be generated on an ongoing basis. As a result, we never know the true Process Standard Deviation, σ. So, we use an estimate which is denoted by $\hat{\sigma}$ ("sigma hat").

This gives us a first cut at a Statistic for Process Capability, the Capability Index known as C_p.

$$C_p = \frac{\text{USL} - \text{LSL}}{6\hat{\sigma}}$$

(Since Capability Indices are usually calculated by just entering the data into a software program, we won't go into the formulas (there are two possibilities) for calculating $\hat{\sigma}$. They involve the use of constants which vary by Sample Size and are usually shown in tables.)

C_p is useful for gaining a conceptual understanding of Capability Indices, but it has serious practical limitations. It assumes that there is both an Upper and Lower Spec limit. Often there is only one or the other. It also assumes that the Process Mean is exactly halfway between the LSL and the USL, which is often not the case.

The **Capability Index C_{pk}** was developed to avoid these limitations.

– It only uses the Spec Limit (USL or LSL) which is closest to the Mean of the data.

- So, instead of comparing a spread of 6σ to the difference from top to bottom of the Spec Limits, it compares 3σ to the spread from the Mean to the closer Spec Limit.

$$C_{pk} = \frac{\textbf{Absolute Value [Mean − Closer Limit]}}{3\hat{\sigma}}$$

In a Normal Distribution 99.7% of the data lie within three Standard Deviations (Sigmas) of the Mean. So, if 99.7% of our data lie between the Spec Limits, then the Spec Limits correspond to 3σ. Given the formula above, C_{pk} would equal 1.

Number of σ's that fit within the Spec Limit	C_{pk}
3	1.0
4	1.33
5	1.66
6	2.0

C_{pk} **is the most effective and most widely used Capability Index**. It assumes the data are Normally Distributed, which is how random Variation within a Process which is Under Control would be distributed.

There are other Capability Indices, including one for individual machines and one when target values are used rather than Spec Limits.

Capability Indices measure short-term Variation – the Variation observed when a Sample of **data is collected in a short period of time under essentially the same conditions**. Such a Sample is called a Rational Subgroup.

 | **4. For every Capability Index, there is a corresponding Performance Index which measures how well the Process actually performs over the long term.**

One might expect that, over longer periods of time, Variation would increase. The Process Mean might "drift" from its initial value, or the Standard Deviation might increase – or both. Conditions can change, and the Process may no longer be stable and Under Control.

Process Performance Indices are intended to measure this. The data are collected in subgroups over a period of time. From each subgroup, a Capability Index could be calculated. If all the data in all the subgroups are lumped together and treated as a single Sample, then a Performance Index can be calculated. The formulas for Capability and Performances are the same; the difference is the length of time over which the data are collected.

	Capability Index	**Performance Index**
Example index	C_{pk}	P_{pk}
Data collected	in subgroups	ungrouped
Measures	within-group Variation	Variation across all the data of multiple subgroups (includes between-group Variation)
	short-term Variation	long-term Variation
	the best the process is **capable** of	how the process actually **performs** over the long term

 5. A Capability Index can be used to calculate the Sigma quality level of a Process and to guide efforts to improve the Process. **There is, traditionally, an assumed 1.5 Sigma drop from Capability to Performance.**

We showed earlier how a value of C_{pk} can be calculated from the number of Standard Deviations (Sigmas) which fit within the Specification Limits. And we know from the Empirical Rule (see the article *Normal Distribution*) exactly what percentage of data points fall within any number of Standard Deviations. For example,

- 68.27% within one Standard Deviation
- 95.45% within two Standard Deviations
- 99.73% within three Standard Deviations

Likewise, we know the percentages for four, five, six, and more Sigmas. This enables us to convert values of C_{pk} and P_{pk} into Sigmas.

In the "Six Sigma" (and "Lean Six Sigma") process improvement disciplines, "Sigmas" are used as a shorthand to describe the quality of the Process – as measured by the number of DPMO for a defect.

To account for the difference between short-term and long-term Variation, the convention in Six Sigma is to claim a DPMO rating corresponding to 1.5 Sigma less than that measured in a short-term study.

The objective, as determined by data in a short-term study, is to have six Standard Deviations (Sigmas) fit within the Spec Limits. If that is achieved, a Process can claim 3.4 DPMO. That is, only 3.4 data points out of one million are outside the Spec Limits. But, for a Normal Distribution, that is what you get when 4.5 Standard Deviations are inside the Spec Limits.

C_{pk}	Sigma Level	DPMO	% Good
0.33	1	691,463	31%
0.67	2	308,537	69.1%
1.00	3	66,807	93.3%
1.33	4	6,210	99.4%
1.67	5	233	99.977%
2.00	6	3.4	99.99966%

Related Articles in This Book: *Control Charts – Part 1: General Concepts and Principles*; *Standard Deviation*; *Normal Distribution*

PROPORTION

Summary of Keys to Understanding

1. **Proportion is the primary Statistic used for summarizing Count data from Categorical** (aka Nominal) **Variables. It is the decimal equivalent of a percentage.**

Categorical Variable	gender	
categories	female	male
data: Counts	44	36
Statistic: Proportion = Count/Total	44 / 80 = 0.55	36 / 80 = 0.45

2. **The Count data in a 2-category Variable generally follow a Binomial Distribution.**

3. The Central Limit Theorem (CLT) applies to Proportions as well as to Means: **For a Proportion p, provided that np > 5 and $n(1 - p) > 5$,**

   the **Distribution of the Sample Proportions will be Normal, with**

   the **Mean of the Sampling Distribution** $= p$

   the **Standard Deviation of the Sampling Distribution** $= \sqrt{p(1-p)/n} = \sqrt{pq/n}$

4. **Therefore, the Test Statistic, z, can be used to solve problems involving Proportions of 2-category Count data, for example,**
   - **Estimate the Population or Process Proportion.**
   - **Is there a Statistically Significant difference between two Population Proportions?**

5. **The Chi-Square test for Independence can be used for Proportions of Count data in three or more categories.**

Explanation

> **1. Proportion is the primary Statistic used for summarizing Count data from Categorical** (aka Nominal) **Variables. It is the decimal equivalent of a percentage.**

A Statistic is a numerical property of Sample data, for example, the Mean or Standard Deviation. "Parameter" is the name for the corresponding property of the Population or Process from which the Sample was drawn.

Statistics and their corresponding Parameters usually have different symbols. Mean and Standard Deviation use Greek letters for their Parameters and Roman (the English alphabet) for Statistics.

p **is the symbol for the Proportion of a Population or Process.**

p **is also the symbol for Probability.** The two concepts are closely related. If the Proportion of green candies in a bin of holiday candies is 0.35, then the Probability of blindly picking a green candy is 0.35.

$\hat{p}$ ("p-hat") **is the symbol for the Proportion of a Sample.** The hat is usually used for estimates, and Statistics from Samples are, by definition, estimates of their corresponding Parameters.

Proportion is a Statistic used with Count data from Categorical Variables. Count data consists of non-negative integers, e.g., 0, 1, 2, 3, etc.

A Categorical Variable (also known as a Nominal or Attributes Variable) divides data into two or more categories. The values of the Variable are names (hence Nominal) of the categories.

Categorical Variable	gender	
Values of the Variable (category names)	female	male

Contrast this with a Continuous or Measurement type of Variable. For the Variable height, the values of the Variable could be Measurement data in inches, e.g., 65.2, 72.6, 70.4, 68.9, etc.

For Categorical/Nominal Variables, the data are the Counts in each category.

In the example below, the data are the Counts of females and males from a random Sample of 80 people entering a building. 80 is the Total.

Categorical Variable	gender	
categories	female	male
data: Counts	44	36
Statistic: Proportion	44 / 80 = 0.55	36 / 80 = 0.45

Proportion = Count / Total

Proportion is expressed as a decimal. The Statistics are the two Proportions – 0.55 for category named "female" and 0.45 for the category named "male." Note that the Proportions are just the decimal equivalent of the two percentages (55% and 45%).

This being statistics, more than one term can be used for the same thing. "Frequency" is sometimes used instead of "Count." "Relative Frequency" is sometimes used instead of Percentage. (Percentage is Proportion multiplied by 100).

> **2. The Count data in a 2-category Categorical Variable generally follow a Binomial Distribution.**

If the following four conditions are met, Samples of Counts will follow a Binomial Distribution. (See the article by that name for more information.)

- Each Sample has the same number of "trials" (e.g., a coin flip or an inspection of an item).
- Each trial can have only one of two outcomes. This means that there are only two categories in the Categorical Variable, for example,

Categorical Variable	categories (values) of the Variable
gender	female, male
candy color	green, not green
item inspection	defective, not defective

- The trials are Independent and do not affect one another.
- The Probability of a success is the same for all trials.

  The term "success" indicates that the item selected has the property we are interested in counting. For example, if we are interested in the Proportion of green candies, a trial which selected a candy that turned out to be green would count as a success. (Somewhat perversely, if we were inspecting for items that had defects, a defective item would be called a "success.")

There are many Binomial Distributions. Much as a unique Normal Distribution can be identified by specifying the Mean and Standard Deviation, **a unique Binomial Distribution can be described by specifying the Proportion (p) of successes and the Sample Size (n).**

The Proportion determines whether the tail of a Binomial Distribution will be skewed to the right ($p < 0.5$), to the left ($p > 0.5$), or whether the Distribution will be symmetrical ($p = 0.05$).

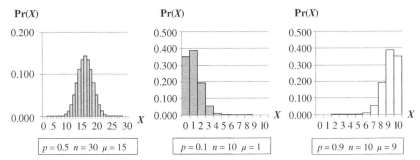

The horizontal (*X*) axis shows the possible Counts of successes in a Sample. For *n* = 10, the possible Counts are 0 through 10. For *n* = 30, the Counts are 0 through 30. **The vertical axis shows the Probability of getting that Count in any given Sample.**

As, we'll see next, **whereas the Probability of any given value of a Count can be determined using the appropriate Binomial Distribution, the Probability of any given value for Proportion can be determined using a Normal Distribution.**

 3. The Central Limit Theorem (CLT) applies to Proportions as well as to Means: **For a Proportion *p*, provided that $np > 5$ and $n(1 - p) > 5$,**

the **Distribution of the Sample Proportions will be Normal, with**

the **Mean of the Sampling Distribution** $= p$

the **Standard Deviation of the Sampling Distribution** $= \sqrt{p(1 - p)/n} = \sqrt{pq/n}$

"provided that np > 5 and $n(1 - p) > 5$" is sometimes replaced with the statement "provided the Sample is large enough." And then, "np > 5 and $n(1 - p) > 5$" is given as the definition of "large enough."

Below are some values of *p* and the minimum Sample Size ("min. *n*") to get np and $n(1 - p) > 5$. You can see that only the smaller of the two (*p* or $1 - p$), is the deciding value. These are shaded.

You can also see that larger Sample Sizes are needed when the Proportions are very different (0.9 and 0.1) and the smallest Sample size corresponds to the Proportions being equal at 0.5.

p	0.1	0.2	0.3	0.4	0.5	0.6	0.7	0.8	0.9
np	*5.1*	*5.2*	*5.1*	*5.2*	*5.5*	7.8	11.9	20.8	45.9
min. *n*	**51**	**26**	**17**	**13**	**11**	**13**	**17**	**26**	**51**
$1 - p$	0.9	0.8	0.7	0.6	0.5	0.4	0.3	0.2	0.1
$n(1 - p)$	45.9	20.8	*11.9*	7.8	*5.5*	*5.2*	*5.1*	*5.2*	*5.1*

But this isn't the only consideration which determines the minimum Sample Size. (See the article *Sample Size – Part 1: Proportions for Count Data.*) Those other considerations usually result in minimum Sample Sizes larger than those needed to support the CLT minimum.

The article *Normal Distribution*, describes how taking the Mean of a Sample has a smoothing effect on the Distribution. The Proportion has a similar effect. So, **no matter the shape of the raw data, the Distribution of all possible Sample Proportions of a given size n (a Sampling Distribution) will be a smooth, symmetrical Normal Distribution.** This is called the Central Limit Theorem (CLT).

The Mean of the Sampling Distribution of Proportions is the Population Proportion, p.

Its Standard Deviation is $\sqrt{p(1-p)/n}$ <u>Notation</u>: some publications use q to represent $1-p$.

- **the larger the sample Size, n, the smaller the Standard Deviation**
- **The closer p (and therefore $1-p$) is to 0.5, the larger the Standard Deviation**

It stands to reason that the larger the Sample Size, the more accurate the Sample would be as an estimate of the Population or Process. So its Variation as measured by the Standard Deviation would be smaller. Having $\sqrt{n}$ in the denominator ensures that.

But, what about the numerator, $\sqrt{p(1-p)}$?

p	0.1	0.2	0.3	0.4	0.5	0.6	0.7	0.8	0.9
$1-p$	0.9	0.8	0.7	0.6	0.5	0.4	0.3	0.2	0.1
$p(1-p)$	0.09	0.16	0.21	0.24	0.25	0.24	0.21	0.16	0.09

We can see that **the largest values for** the numerator (and thus for **the Standard Deviation) are produced when p** (and, therefore, $1-p$ also) **is closest to 0.5.** This also makes intuitive sense. Let's say we take a Sample of 100 and have a count of 50 ($p = 0.5$). Plus or minus 20% from 50 would include a range from 40 – 60. If, on the other hand, we have a count of 5 ($p = 0.5$), plus or minus 20% covers a range from 4 to 6.

For more of an understanding of the Central Limit Theorem, read the article, *Normal Distribution*.

> **4. Therefore, the Test Statistic, z, can be used to solve problems involving Proportions of 2-category Count data, for example,**
>
> - **Estimate the Population or Process Proportion.**
> - **Is there a Statistically Significant difference between two Population Proportions?**

z is the Test Statistic used with data that follow a Normal Distribution. A Test Statistic is one that has an associated Probability Distribution from which we can calculate Probabilities for values of the Test Statistic.

Example #1: Confidence Interval Estimate of a Process Proportion

We use the formula

$$p = \hat{p} \pm z\sqrt{s/n}$$

p is the Population or Process Proportion and is $\hat{p}$ the Sample Proportion

z is the value of the Test Statistic z for a specified Level of Significance, Alpha

s is the Sample Standard Deviation and n is the Sample Size

What is the Proportion of defective units produced by a Process? (Note: this method could also be used for an estimate of the Proportion of a Population.)

Before we collect the data, we must select a value for Alpha (α), which is the highest Probability of an Alpha Error that we will tolerate. We want a 95% Confidence Level, so we select $\alpha = 5\%$. We then conduct an inspection of 200 units and find that 8 were defective.

So, the Sample Proportion $\hat{p} = 8/200 = 0.04$. We don't know the Process Proportion, so we use the Sample Proportion $\hat{p}$ to determine whether the Sample is large enough.

$n\hat{p} = (200)(.04) = 8$ and $n(1-\hat{p}) = (200)(.96) = 192$. Both of these are more than five, so we can proceed.

Our estimate of the Process Proportion (p) is the Sample Proportion, 0.04, plus or minus a Margin of Error (MOE).

$$p = 0.04 \pm \text{MOE}$$

We can use z to calculate the Confidence Interval depicted above – that is, the Range around p which includes the MOE. The formula for MOE is

$$\textbf{MOE} = z\sqrt{s/n}$$

We look up the value of z for $\alpha = 5\%$ and a 2-tailed analysis (MOE includes plus and minus) and find that $z = 1.96$.

We also need to know the value of s, the Sample Standard Deviation. We use the formula for the Standard Deviation of the Sampling Distribution:

$$s = \sqrt{\hat{p}(1-\hat{p})/n} = \sqrt{(0.04)(0.96)/200} = 0.014 \text{ so}$$
$$\text{MOE} = 1.96\sqrt{0.014/200} = 0.016$$
$$\boldsymbol{p = 0.04 \pm 0.016}$$

So, with a Confidence Level of 95%, we can say that the Process Proportion is 0.04 plus or minus 0.016

Note: The formula for MOE above can be algebraically manipulated to become a formula for n, the Sample Size required to achieve a given value of MOE. See the article *Sample Size – Part 1: Proportions for Count Data* for details.

Example #2: Hypothesis Test of the difference in Proportions between two Populations

We want to know whether there is a Statistically Significant difference between the Population of female voters and the Population of male voters in their preference for a particular candidate for public office.

Before collecting data, we select $\alpha = 5\%$. For a 2-sided test, this gives us the Critical Value of z; z-critical $= 1.96$.

The Null Hypothesis, H_0, states that there is no difference. We'll do a 2-tailed analysis, since we don't care whether the difference is on one direction or the other, just whether there is a difference.

We then surveyed 100 women and find that 52 prefer the candidate. We surveyed 80 men and 38 said they preferred the candidate.

So, the Counts are $X_1 = 52$ and $X_2 = 38$. And the Proportions are $\hat{p}_1 = 0.52$ and $\hat{p}_2 = 38/80 = 0.475$. Is this a Statistically Significant difference?

For this type of test, the Samples are large enough if there are at least five successes and five failures in each of the Populations. That is the case here.

The formulas get somewhat complicated, so software is usually used. The result:

$z = 0.60$, so $z \leq z$-critical.

The p-value is 0.548, so $p > \alpha$.

(Note that this p is not a Proportion. It is the Probability of an Alpha Error.)

Both of these results tell us the same story: So we Fail to Reject (we Accept) the Null Hypothesis of no difference. There is not a Statistically Significant difference.

This 2-tailed z-test for two Population Proportions is statistically the same as a Chi-Square Test for the Independence using a 2 by 2 table. (2 Populations, 2 Proportions).

 5. The Chi-Square test for Independence can be used for Proportions of Count data in three or more categories.

Juice Study: Proportions are the same, so the Variables Gender and Juice are Independent						Ice Cream Study: Proportions are very different, so the Variables Gender and Ice Cream are Associated (not Independent).				
	female		male				female		male	
	Count	Proportion	Count	Proportion			Count	Proportion	Count	Proportion
apple	28	0.35	14	0.35		chocolate	48	0.48	16	0.20
grape	12	0.15	6	0.15		strawberry	28	0.28	40	0.50
orange	40	0.50	20	0.50		vanilla	24	0.24	24	0.30
Total	80	1.00	40	1.00		Total	100	1.00	80	1.00

There are three commonly used Chi-Square tests – for Goodness of Fit, Independence, and Variation. This book has articles for each of them. All three tests analyze Count data. The Chi-Square Test for Independence focuses on the Proportion Statistic.

If the Proportions are the same (or there is no Statistically Significant difference) then the Variables are said to be Independent. In our Example #2 above, there was no Statistically Significant difference, so the Variables gender and candidate preference were Independent.

In the example of the ice cream study, illustrated in the right table above, the Variables gender and Ice Cream preference are <u>not</u> Independent. According to this data, person's gender <u>does</u> affect their choice in Ice Cream flavor to a Statistically Significant degree.

Related Articles in This Book: *Sample Size – Part 1: Proportions/ Percentages*; *Binomial Distribution*; *Normal Distribution*; *Sampling Distribution*; *Standard Deviation*; *Test Statistic*; *z*; *Alpha, a*; *Statistically Significant*; *Confidence Intervals – Parts 1 and 2*; *Margin of Error*; *Hypothesis Testing – Parts 1 and 2*; *Null Hypothesis*; *Fail to Reject the Null Hypothesis*; *Chi-Square Test for Independence*

r, MULTIPLE R, r^2, R^2, R SQUARE, R^2 ADJUSTED

This article is for clarifying possible confusion about similar terms. For details, see the articles: Correlation – Part 2; Regression – Part 2: Simple Linear; and Regression Part 4: Multiple Linear.

Summary of Keys to Understanding

Term	Name	Used In	Comments
r	Correlation Coefficient	Correlation	• Values range from -1 to 1. • Values near -1 or 1 indicate strong Negative or strong Positive Correlation. • 0 indicates no Correlation.
Multiple R	Multiple Correlation Coefficient	Multiple Linear Regression	When there are only two Variables, Multiple $R = r$.
r^2			It's just the square of r. In Simple Linear Regression, $r^2 = R^2$.
R^2	R Square aka Coefficient of Determination aka Squared Error	Simple Linear Regression and Multiple Linear Regression	The most common measure for the Goodness of Fit of a Regression line. • Values range from 0 to 1. • Values nearer to 1 indicate a better fit.
$\bar{R}^2$	Adjusted R^2 aka R^2 Adjusted	Simple Linear Regression and Multiple Linear Regression	• Adjusts R^2 for Degrees of Freedom. • Experts disagree on whether it is necessary.

Explanation

All these terms are used in Correlation analysis and Linear Regression.

Term	Name	Used In	Comments
r	Correlation Coefficient	Correlation	• Values range from -1 to 1. • Values near -1 or 1 indicate strong Negative or strong Positive Correlation. • 0 indicates no Correlation.

r is the Correlation Coefficient. (See the article *Correlation – Part 2*.) It is a measure of the linear Correlation between two Variables, *x* and *y*. *r* is also known as the "Pearson product-moment correlation coefficient," "PPMCC" or "PCC," or "Pearson's *r*." *r* ranges in value from -1 to $+1$.

If $r = 0$, there is no Correlation. The closer *r* is to -1 or 1, the stronger the Correlation. Negative values indicate Negative Correlation – as one Variable increases in value, the other decreases. Positive values indicate Positive Correlation – both Variables increase together.

Term	Name	Used In	Comments
Multiple *R*	"Multiple Correlation Coefficient"	Multiple Linear Regression	• Values range from -1 to 1. • Values near -1 or 1 indicate strong Negative or strong Positive Correlation. • 0 indicates no Correlation. When there are only two Variables, Multiple $R = r$.

Multiple *R* is the "Multiple Correlation Coefficient." It measures the Correlation between the values of *y* (the Dependent Variable) predicted by the Regression Model and the actual *y*'s in the data. So, it is a measure of the Goodness of Fit of the Regression Model.

$$-1 \leq \text{Multiple } R \leq 1$$

Similar to *r*, Multiple $R = 0$ indicates no Correlation. The closer Multiple *R* is to -1 or $+1$, the stronger the Correlation.

The main purpose of Multiple *R* appears to be as an interim calculation for R^2.

Term	Name	Used In	Comments
r²			It's just the square of *r*. In Simple Linear Regression, $r^2 = R^2$.

Term	Name	Used In	Comments
R^2	*R* Square aka Coefficient of Determination aka Squared Error	Simple Linear Regression and Multiple Linear Regression	The square of Multiple *R*. The most common measure for the Goodness of Fit of a Regression line. • Values range from 0 to 1. • Values nearer to 1 indicate a better fit.

R^2 **is a measure of how well the Regression Model fits the data.** It is the portion of the total Variation in *y* which is explained by the Model.

Being the square of Multiple *R*, which ranges from −1 to 1, **the range of R^2 is: $0 \leq R^2 \leq 1$.**

Higher values indicate a better fit of the Regression Model to the data.

R^2 has a shortcoming: As the number of *x* Variables increases, the value of R^2 also increases, irrespective of how good the fit is. That is the reason for Adjusted R^2. (See the article *Regression – Part 4: Multiple Linear.*)

Term	Name	Used In	Comments
$\bar{R}^2$	Adjusted R^2 aka R^2 Adjusted	Simple Linear Regression and Multiple Linear Regression	• Adjusts R^2 for the number of *x* Variables. • Experts disagree on whether it is necessary.

The purpose of the adjustment is to counteract the shortcoming in R^2 noted above. However, not all experts agree it is needed.

Related Articles in This Book: *Correlation – Part 2*; *Regression – Part 2: Simple Linear*; *Regression Part 4: Multiple Linear*; *Degrees of Freedom*

REGRESSION – PART 1 (OF 5): SUMS OF SQUARES

Summary of Keys to Understanding

1. **The Sum of the Squared Deviations – from each value of y to the Mean of y – is called the Sum of Squares Total, or SST.**
   SST measures ALL the Variation in the Variable y around the Mean of y.

2. In Regression, the **Sum of Squares Error, SSE, measures the Variation in y which IS NOT explained by a Regression Line**. It is the Sum of the Squared Deviations from each value of y to a Regression Line. **It is a component of SST.**

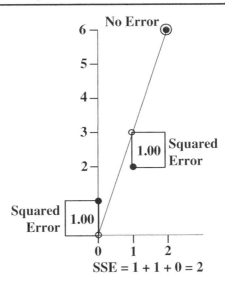

$$\text{SSE} = 1 + 1 + 0 = 2$$

3. **Sum of Squares Regression (SSR) is that part of SST which IS modeled by a Regression Line.** And, since SST = SSE + SSR,
   $$\text{SSR} = \text{SST} - \text{SSE}.$$
   $R^2 = \text{SSR/SST}$ is a measure of the Goodness of Fit of the Regression Line.

4. **The Best Fit line is the one with the smallest SSE – the "(Ordinary) Least Sum of Squares."**

Explanation

> **1. The Sum of the Squared Deviations – from each value of y to the Mean of y – is called the Sum of Squares Total, or SST.**
>
> $$\text{SST} = \Sigma(y - \bar{y})^2$$
>
> **SST measures ALL the Variation in the Variable y.**

If we have a Sample or other group of data values for a single Variable, call it y, we can calculate the Mean, $\bar{y}$. Then **Deviation** (from the Mean) of any single data value, y is defined as $y - \bar{y}$.

Example 1: one Variable (y)

Let's say we have a Sample of three values of a single Variable y: 1, 2, and 6. These are plotted along the vertical (y) axis in the diagram below.

The Mean is $\bar{y} = (1 + 2 + 6) / 3 = 3$.

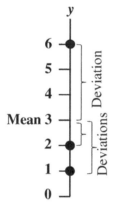

Mean $\bar{y} = 3$	
y	Deviation $y - \bar{y}$
1	$1 - 3 = -2$
2	$2 - 3 = -1$
6	$6 - 3 = 3$

If we want to get a measure of how much Variation (also known as "Variability," "Spread," or "Dispersion") there is in the Sample, we cannot just total up the Deviations for each of the data values. That is because the data points with negative values for Deviation from the Mean would cancel out those with positive Deviations, resulting in a total Deviation of zero. (This is guaranteed by the definition of Mean.)

The **Squared Deviation** of any single data value of y is $(y - \bar{y})^2$. Squaring eliminates the negative values. It also disproportionately increases the impact of data points farther from the Mean. (3^2 is more than 3×1^2.) This is consistent with how things are done in calculating the Standard Deviation and Variance.

The Sum of the Squared Deviations from the Mean is called the "Sum of Squares Total," or SST.

$$\text{Sum of Squares Total: SST} = \Sigma(y - \bar{y}^2)$$

If we draw squares with sides which are the length of the Deviations, the areas of the squares are equal to the Squared Deviations:

Example 1 continued: one Variable (y)

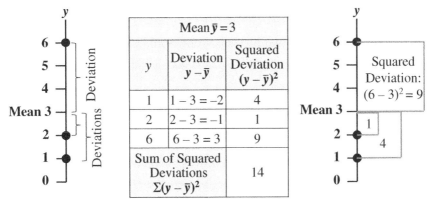

In the illustration above, SST = $4 + 1 + 9 = 14$.

The Sum of Squares Total includes all of the Variation in the values of the y Variable. There is no other Variation in the y Variable that is not included in SST.

---

 2. In Linear Regression, the **Sum of Squares Error (SSE)** measures the **Variation in y which is NOT explained by the Regression Line**. It is the Sum of the Squared Deviations from each value of y to a Regression Line. **It is a component of SST.**

When all we have is the single Variable y, then all we have is a 1-dimensional (y-axis only) graphs like those above, and that is the end of the story.

But what if we suspect that the Variation in y is caused by another Variable x? Then, we'll need to collect data in (x, y) pairs, e.g., x = years of education and y = income for a number of individuals.

In Simple Linear Regression, we try to fit a line to paired (x, y) data.

The "Error" in Sum of Squares Error is the error in the Regression Line as a Model for explaining the data.

- SST is the Sum of Squared Deviations from the Mean.
- **SSE** is the **Sum of Squared Deviations from the Regression Line.**

$$\text{SSE} = \Sigma(y - y_{\text{LINE}})^2$$

- SST is the total of all Variation in the y Variable.
- **SSE** is: the **part of SST not explained by the Regression Line.**

$$\text{SSE} = \text{SST} - \text{SSR}$$

To illustrate, let's say that instead of our y-only data values of 1, 2, and 6, we collected data in (x, y) pairs: $(0, 1)$, $(1, 2)$, and $(2, 6)$.

In the table below, the first two columns show the paired (x, y) data. If we plot the data, we might take a guess that a line from $(0, 0)$ to $(2, 6)$ might be a good start in trying to fit a line to the data. The equation for that line is $\hat{y} = 3x$. ($\hat{y}$, "y hat," is the predicted value for y, given our first estimate of a Regression line.) Using this equation, we can calculate the values of y on the line for $x = 0$, 1, and 2. Then, we calculate the Deviation and the Squared Deviation.

Example 2: paired (x, y) data and an **estimated** Regression Line, $\hat{y} = 3x$

x	y		$\hat{y}$	**Deviation $y - \hat{y}$**	**Squared Deviation (Squared Error)**
0	1		0	1	1
1	2		3	-1	1
2	6		6	0	0
Sum of Squared Errors (SSE) = Sum of Squared Deviations					2

Several things should be noted at this point:

- The equation $\hat{y} = 3x$ is just our first estimate of a Regression Line for purposes of illustration. We don't know if it is a good fit to the data or not.
- **A Regression Line (or curve) is described by a Model** (which is a formula like $\hat{y} = 3x$). **Given a value for x, the Model will predict a corresponding value for y.**
- **A Deviation from the Regression Line or curve is called an Error.** This is because **the values of y on a Regression Line or curve are predictions** made by the Model. And a difference between a

prediction and the actual data is considered to be an Error in the Model.

The diagrams below display the information in the table – Example 2 – above. The three (x, y) data points are shown as black dots. The circles are $(x, \hat{y})$ points on our estimated Regression Line. Each of these points has the same x-value as one of the data points, but the $\hat{y}$ values of these points are calculated using the equation for the estimated Regression Line $\hat{y} = 3x$.

The **Deviations** between the y-values of the black data dots and the $\hat{y}$ values of the circles represent the **Error** in the estimated Regression line as a Model for explaining the Variation in the y Variable in the data.

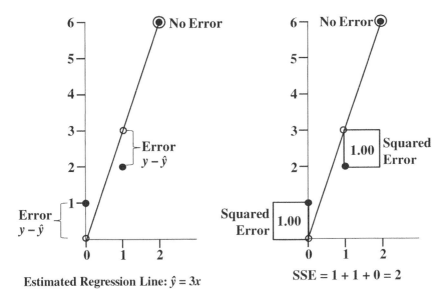

Estimated Regression Line: $\hat{y} = 3x$

SSE = 1 + 1 + 0 = 2

In the diagram above right, the **Squared Errors** are illustrated as squares.

Sum of Squares Error (SSE) is the sum of the Squared Errors from the Regression Line.

---

3. **Sum of Squares Regression (<u>SSR</u>) is <u>that part of SST which IS modeled by a Regression Line</u>.** And, since SST = SSE + SSR,

$$SSR = SST - SSE$$

R^2 = **SSR/SST is a measure of the Goodness of Fit of the Regression Line.**

Given what we've shown above, it is a very simple step to determine how much of the Variation in the Variable y is modeled by a Regression Line:

- SST is ALL the Variation in the Variable y.
- SSE is the Variation in y, NOT modeled by a Regression Line.
- So, the Total (SST) minus what's NOT (SSE) is what IS (modeled by a Regression Line). This is:

Sum of Squares Regression: SSR = SST − SSE

In our example with an estimated Regression Line, SST = 14 and SSE = 2, so SSR = 12.

Is this a good fit? A good fit would have a large value of SSR (which means, a small value of SSE). But simply a large value of SSR can be misleading, because SST might be very large. Obviously, we're looking for a large value of SSR relative to the value of SST. **So SSR/SST would be an appropriate measure of Goodness of Fit** of the Regression line to the data.

SSR/SST is denoted by R^2 and is called the "Coefficient of Determination." More on that in the Part 2 article.

 | **4. The Best Fit line is the one with the smallest SSE – the "(Ordinary) Least Sum of Squares."**

There are several methods for determining the Best Fit, but the most common method is to calculate Sums of Squares, as we have above, and then to determine **the Least Sum of Squares** – the lowest SSE. (This is also known as OLS, the Ordinary Least Sum of squares.)

Since SSE = SST – SSR, **the lowest SSE corresponds to the highest SSR.** And R^2 = SSR/SST is a measure of Goodness of Fit. So the Best Fit could just as easily been described as the one with the highest value of SSR or the highest value of R^2.

Spreadsheets or statistical software can calculate the Best Fit Regression Line for you. In our example, the Best Fit line (the Regression Model) was described by the equation: $y = 0.5 + 2.5x$. This line gives us the (x, y) points $(0, 0.5)$, $(1, 3)$, and $(2, 5.5)$.

$$SSE = 0.25 + 1.00 + 0.25 = 1.5$$

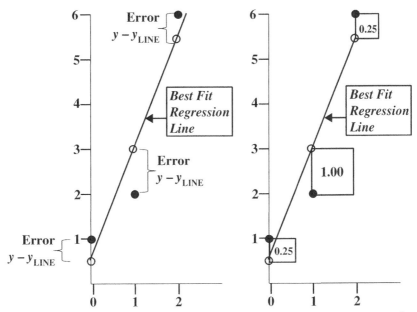

Its SSE is 1.5 (compared to 2 for our estimated line). And its $R^2 =$ SSR/SST $= 14.5/16 = 90.6\%$, compared to $14/16 = 87.5\%$. (We were pretty lucky with our estimated line being such a good fit. But it only had three data points, and we're not going to see that too often in actual practice.)

A few other items before we go:

- We'll elaborate on this in the Part 3 article, but **don't extrapolate the Regression Line or Curve beyond the Range of the x's in the data.** The Model is <u>only</u> valid within that Range – in this example between $x = 0$ and $x = 2$.

- We've focused on a 2-dimensional Regression Line in this article. That is but one type of Regression, **Simple Linear Regression**. **Non-linear Regression**, covered in the *Part 5* article, is not restricted to fitting lines but can use curves of various types.

- Also, if we just have one Independent Variable x, we have **Simple Regression** *(Part 2),* and we can graph it on a simple x-y axis in two dimensions. **Multiple Regression** *(Part 4)* is not Simple, but involves multiple x's and one y. This effectively involves working in three or more dimensions.

- This article deals with the types of Sums of Squares which are used in Regression. Another article, *Sums of Squares*, describes, SS's used with ANOVA and other types of analysis.

For both ANOVA and Regression, Sums of Squares are used to "partition" (allocate, or "divide up") the total Variation (SST) into components which are other types of Sums of Squares.

Related Articles in This Book: *Variation/Variability/Dispersion/Spread*; *Regression – Part 2: Simple Linear*; *Regression – Part 3: Analysis Basics*; *Regression Part 4: Multiple Linear*; *Regression Part 5: Simple Nonlinear*; *Sums of Squares*; *r, Multiple R, r^2, R^2, R Square, Adjusted R^2*

REGRESSION – PART 2 (OF 5): SIMPLE LINEAR

Builds on the content in the article, "Regression – Part 1: Sum of Squares".

Summary of Keys to Understanding

 1. The purpose of Regression analysis is to develop a <u>Cause and Effect</u> "Model" in the form of an equation: $y = f(x)$ or $y = f(x_1, x_2, \ldots, x_k)$.

The Model <u>predicts</u> what **future** (yet-to-be-collected) **data** will be like. **It is validated – or invalidated – in Designed Experiments.**

2. First, plot the data. Then, perform a Correlation analysis. Only perform a Simple Linear Regression if there is a moderately strong Correlation between the Variables.

3. Simple Linear Regression fits a line

$$y = bx + a$$

to 2-dimensional (x, y) data.

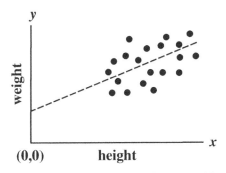

4. The (best fit) **Regression Line is the line with the smallest value of the Sum of Squares Error(SSE).**

5. The Coefficient of Determination, R^2, is a measure of how well the Regression Model fits the data.

$$R^2 = \frac{\text{SSR}}{\text{SST}} = 1 - \frac{\text{SSE}}{\text{SST}} = r^2$$

It is the Proportion of the total Variation in y which is explained by the Regression Model.

Explanation

> **1. The purpose of Regression analysis is to develop a <u>Cause</u> <u>and Effect</u> "Model" in the form of an equation: $y = f(x)$ or $y = f(x_1, x_2, \ldots, x_k)$.**
>
> The Model **predicts** what **future** (yet-to-be-collected) **data** will be like. **It is validated – or invalidated – in Designed Experiments.**

In Correlation analysis, there is no Cause and Effect. Correlation analysis studies whether two Variables are interrelated such that they vary together. The two Variables are considered equal in the sense that we don't consider whether one influences the other. There may also be other unknown Variables involved which influence one or both.

Regression analysis extends Correlation analysis to determine one or more Causes – one or more x variables – **and their Effect** on the y Variable. So unlike Correlation analysis, Regression analysis differentiates between the x and y variables. **Regression is not symmetric.** That is, modeling x vs. y is different from modelling y vs. x.

So, we have different names for the x and the y variables. Also – not surprisingly for statistics – there are several different names for each. Here, we'll use Predictor for x and Outcome for y, since those seem most descriptive

<u>Synonyms:</u>

x	Independent Variable	Cause	Predictor Variable		Explanatory Variable
y	Dependent Variable	Effect	Outcome Variable	Response Variable	Criterion Variable

There is always only one y (Outcome) Variable. It is a function of one or more x's.

Simple Regression: one x (Predictor) Variable: $y = f(x)$

Multiple Regression: multiple x variables: $y = f(x_1, x_2, \ldots, x_n)$

These equations are called Models, because they attempt to model the real-world phenomena described by the data.

If the Model is accurate, it should accurately predict future behavior of the phenomena, as evidenced by any data we collect in the future.

We determine whether the Model is accurate using the statistical discipline of Design of Experiments (DOE). We test various levels of the x Variable(s) and predict the resulting values of the y Variable. The results of the Designed Experiments determine whether or not the Model is valid.

Sometimes it's not always clear which Variable should be the x Variable and which would be the y. **In general, the Outcome Variable, y, is the** one we are studying, the **Variable whose value we want to predict,** and **the Predictor Variables, (x's) are used to make that prediction.**

For example, we are interested in the reading level of children entering 1st grade. We believe that a Factor influencing the reading level would be the parents' educational level. The Outcome Variable, y, would be the reading level and one of the Predictor Variables, x's, would be the parents' educational level.

> **2. First, plot the data. Then, perform a Correlation analysis. Only perform a Simple Linear Regression if there is a moderately strong Correlation between the Variables.**

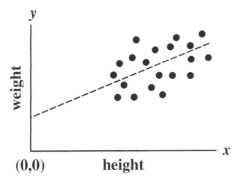

As described in the article, *Correlation Analysis Part 2,* **statistics alone can be misleading. It is important to plot the data.** For Simple (with one x Variable) Regression, data can be plotted in a 2-dimensional Scatterplot. **For Simple Linear Regression** – which has a single x Variable – **the Scatterplot should show at least a roughly linear pattern. If it does not, do not proceed further.**

If it does, the next step is to determine the strength of the linear Correlation. Perform a Correlation analysis. This produces a Correlation Coefficient, r, which is a measure of the strength of a linear relationship between two Variables. If r does not indicate that there is at least a moderately strong correlation ($r \geq 0.50$ or $r \leq -0.50$), it would not be productive to perform a Simple Linear Regression analysis to fit a line and make predictions.

> **3. Simple Linear Regression fits a line to two dimensional (x, y) data.** The equation describing the line is
>
> $$y = bx + a$$

"Simple," as opposed to "Multiple," means there is only one x Variable.

"Linear," as opposed to "Nonlinear," means the equation describes a line.

You may see variations on this equation, such as $y = a + bx$ or $y = ax + b$ or $y = mx + b$ or $y = b_0 + b_1 x$. But it's all the same thing, just different letters and ordering.

The data are "paired" as in Correlation. That is, each data point is of the form (x, y). It consists of one value of x and one value of y from a single test subject – for example, the height and weight of one person. (It makes no sense to pair the height of one person with the weight of another person.)

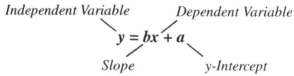

The **Slope**, denoted by the constant b, **describes the slant of the line**. In these diagrams, the line described by the equation is dotted.

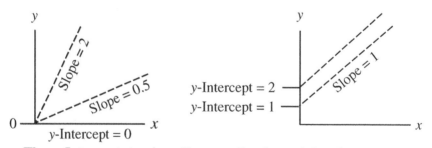

The **y-Intercept** (or just "Intercept"), denoted by the constant a, **describes where the line intercepts the y-axis. It is the value of y for which $x = 0$.**

So, how do we determine the values of b and a, which define the Regression Line? There are a number of methods for doing so. But the most widely used is the Least Squares method. This is described in the Part 1 article. **The Regression Line is the one which minimizes the SSE** (Sum of Squares Error).

The following are the formulas for calculating the slope and intercept of the Regression Line. You won't normally use these, since spreadsheets or software can provide these values. First, calculate the Slope, b:

$$\text{Slope: } b = \frac{\sum(x - \bar{x})(y - \bar{y})}{\sum(x - \bar{x})^2}$$

This may look somewhat reminiscent of the formula's use in Correlation. If fact, with a little algebraic manipulation, it can be rewritten in terms of the Correlation Coefficient, r, and the Standard Deviations of x and y:

$$\text{Slope: } b = r\frac{s_y}{s_x}$$

Having calculated the Slope, the y-Intercept is a simple matter.

$$y\text{-Intercept: } a = \bar{y} - b\bar{x}$$

The y-Intercept does <u>not</u> always have a physical meaning. In the illustration below, we show some data on height and weight of people. The line through the data is extended to the left until it intercepts the y axis (at $x = 0$). This is the point at which the height of a person would be zero which, of course, is meaningless.

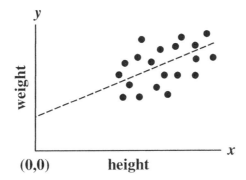

$(0,0)$ **height**

---

 4. The (best fit) Regression Line is the line with the smallest value of the Sum of Squares Error, SSE.

There are a number of methods for calculating a line which best fits the data. The one most commonly used is the Least Squares method. As explained in the Part 1 article, Sum of Squares Error, SSE, is the Variation in y NOT modeled by a line. The line with the smallest value of SSE is the Regression Line. It is also called the Best Fit Line or the Line of Least Squares.

The Regression Line always passes through the point $(\bar{x}, \bar{y})$.

> **5. The Coefficient of Determination, R^2, is a measure of how well the Regression Model fits the data.**
>
> $$R^2 = \frac{\text{SSR}}{\text{SST}} = 1 - \frac{\text{SSE}}{\text{SST}} = r^2$$
>
> It is the Proportion of the total Variation in y which is explained by the Regression Model.

In Part 1, we described how **the total Variation in the Variable y can be expressed as the Sum of the Squares Total, SST**.

As described in Part 2, we can develop a Regression Model – in Linear Regression, this is an equation for a line. The Regression Model, will explain a portion – hopefully a major portion – of the Variation. **The part of the Variation in the Variable y explained by the Regression Model is expressed as the Sum of Squares Regression (SSR)**. (That's why it is sometimes called the "Explained Variation.")

What's left over – the Unexplained Variation – **is** viewed as the Error in the Regression Model, **the Sum of Squares Error, or SSE**. So, ...

$$\text{SST} = \text{SSR} + \text{SSE}$$

Example:
Let's say we have some data and a Regression Model, and we calculate:

$$\text{SST} = 100, \quad \text{SSR} = 80, \quad \text{and} \quad \text{SSE} = 20$$

The Proportion of the total Variation in the Variable y which is explained by the Regression Model is

$$\frac{\text{SSR}}{\text{SST}} = \frac{80}{100} = 0.8 = 80\%$$

The Proportion of the Error in the Regression Model is what's left:

$$\frac{\text{SSE}}{\text{SST}} = 1 - \frac{\text{SSR}}{\text{SST}} = \frac{20}{100} = 0.2 = 20\%$$

Clearly a Model which, for example, explains 80% of the Variation in the Variable y is better than a Model which explains only 60%. So, SSR/SST would seem to be a good measure of the Goodness of Fit of the Regression Model.

$$R^2 = \frac{\text{SSR}}{\text{SST}} = 1 - \frac{\text{SSE}}{\text{SST}}$$

As described in the article, *Correlation, Part 2*, the formula for r^2, the Coefficient of Correlation, is made up of Sums of Squares and Standard

Deviations. The formula for the latter is also based on Sums of Squares. And, with some algebraic manipulation (which we won't go into here) r^2 can be shown to equal SSR/SST.

So, **in Simple Linear Regression, the Coefficient of Determination, R^2, equals the square of the Coefficient of Correlation, r.**

$$R^2 = r^2$$

These are the same thing, and the notations are interchangeable. R^2 is also called Multiple R^2. (If these different r's are starting to get confusing, see the article r, Multiple R, r^2, R^2, R Square, R^2 Adjusted.)

Since it equals SSR/SST, R^2 is **the Proportion of the total Variation in the Variable y which is explained by the Regression Model.** Since $R^2 = r^2$, and since r, the Coefficient of Correlation, ranges from -1 to $+1$,

Proportion: $\mathbf{0 \leq R^2 \leq 1}$

R^2 is a decimal between 0 and 1. To convert it to a percentage, multiply by 100.

Percentage: $\mathbf{0\% \leq R^2 \times 100 \leq 100\%}$

Different disciplines have different standards for what is a good enough "Goodness of Fit," as measured by R^2, as explained in the *Part 3* article. **If the Model does not meet the desired clip level, then a different x Variable or additional x Variables can be tried.**

In Multiple Regression (*Part 4 article*), an Adjusted R^2 is sometimes used.

An R^2 which meets the desired clip level is necessary, but not sufficient. The *Regression – Part* 3 article on *Analysis Basics* explains why. And the 3-part series of articles on *Design of Experiments (DOE)* explains how to verify the predictions made by the Model.

Important: Please continue on to read the Part 3 article, which provides Analysis Basics for Simple Linear as well as other types of Regression.

Related Articles in This Book: *Correlation – Part 2*; *Regression – Part 1: Sum of Squares*; *Charts, Graphs, Plots – Which to Use When*; *r, Multiple R, r^2, R^2, R Square, R^2 Adjusted*; *Regression – Part 3: Analysis Basics*; *Regression – Part 4: Multiple Linear*; *Regression – Part 5: Simple Nonlinear*; *Design of Experiments (DOE) — Parts 1–3*

REGRESSION – PART 3 (OF 5): ANALYSIS BASICS

These basics apply to all types of Regression. Additional considerations for Multiple Linear Regression and Non-linear Regression are covered in Parts 4 and 5.

Summary of Keys to Understanding

> **1. Different disciplines have different standards for what is a good enough "Goodness of Fit," as measured by R^2.**
> **An R^2 which meets the desired clip level is necessary, but not sufficient.**

> **2. The Residual** – for each (x, y) data point – **is the difference between the value predicted by the Model, $\hat{y}$, and the actual value of y.**
> **Analysis of Residuals is essential in assessing the validity of the Model.**

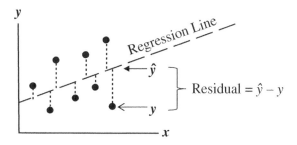

> **3. Cautions:**
> - **Cause and Effect cannot be determined by statistics alone.**
> - **Regression predictions must be tested with experiments on new data.**
> - **Never attempt to predict (extrapolate) beyond the range of the x Variable in the data.**

Explanation

> **1. Different disciplines have different standards for what is a good enough "Goodness of Fit," as measured by R^2.**
>   **An R^2 which meets the desired clip level is necessary, but not sufficient.**

In the **physical sciences and** in **engineering** situations in which precise laws of nature prevail, one might expect $R^2 \geq 0.9$.

In **process improvement** work: $R^2 \geq 0.7$.

In **social sciences, $R^2 \geq 0.3$** may be considered good enough.

If the Model does not meet the desired clip level, then a different x Variable or additional x Variables can be tried.

In Multiple Regression (*Part 4 article*), an **Adjusted R^2** is sometimes used. And statistical software packages can provide p and F values and ANOVA tables to aid in the analysis.

Even if we do have a strong value for R^2, we cannot conclude that our Regression Line is a good fit. **An R^2 which meets the desired clip level is <u>necessary</u>, but <u>not sufficient</u>**, as the rest of this article demonstrates.

> **2. The Residual** – for each (x, y) data point – is the <u>difference</u> between the value <u>predicted</u> by the Model, $\hat{y}$, and the <u>actual</u> value of y.
>   **Analysis of Residuals is essential in assessing the validity of the Model.**

Virtually no Model is a perfect fit for real-world data. For any given value of x (or multiple x's) the predicted (aka "expected") value of y calculated by the Model will usually <u>not</u> be the same as the actual value. This difference is considered an Error in the Model. It is called a Residual.

In the diagram below, the Regression Line is the diagonal dashed line. The black dots are the (x, y) data points. The values of the Residuals are the lengths of the vertical dotted lines. **The Predicted Value of $\hat{y}$ ("y-hat") for a given x is the value of y on the Regression Line for that x.** The actual (aka observed) value of y is that of the data point.

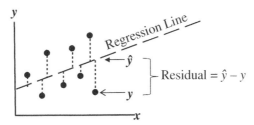

The value of the Residual of any (x, y) data point is the value of y which is predicted (calculated) by the Regression Model (denoted by y with a "hat") minus the actual value of y.

$$\text{Residual} = \hat{y} - y$$

Residuals can contain a lot of useful information, and analysis of Residuals is essential to verifying the accuracy of the Model. The Residuals in any Model should exhibit the following characteristics:

Residuals must:

- be **Random**
- be **Normally Distributed**
- **not** be **Correlated with any** x **Variable**
- **not** be **Autocorrelated** (relative to time sequence)
- have **Constant Variance**
- have **no unexplained Outliers**

To keep this article to a reasonable size, specifics are provided in the article "*Residuals.*"

> **3. Cautions:**
> - **Cause and Effect cannot be determined by statistics alone.**
> - **The Regression Model must be tested with experiments on new data.**
> - **Never predict (extrapolate) beyond the range of the** x **Variable in the data.**

It is important to **engage subject matter experts** to describe credible real-life mechanisms or scenarios by which the Correlation we observe can be explained by x causing y. If it's a mystery why there is a Correlation, then there could be other explanations. For example **an unknown "lurking" Variable could be causing the Variation in both x and y.**

As we said in *Part 2*, the purpose of Regression analysis is to develop a Cause and Effect "Model" in the form of an equation. The Model **predicts** what **future** (yet-to-be-collected) **data** will be like. For Simple Linear Regression, the Model is the equation, $y = bx + a$, which describes a Best Fit line.

A value for R^2 above a commonly accepted clip level and compliant Residuals tells us that we can proceed to the next stage of analysis – a Controlled or Designed Experiment.

We can't use the original data as proof of Cause and Effect. If we want to use the Model to predict values for y, then we must actually predict some new values, and test them with Controlled or Designed Experiments. See the articles *Design of Experiments* (*DOE*) Parts 1–3. So, we must:

- **take a <u>new</u> Sample of data,** and
- **<u>compare</u> the *y*-values predicted by the Model with the Actual *y*-values in the new Sample.**

The experiment should enable us to fix values of *x* at several levels – for example, specific doses of a new drug – and then predict what the corresponding *y*-values will be.

Probably the biggest No-No is extrapolation. Never predict beyond the range of the *x* Variable in the data.

Let's say there is a clinical trial of a new drug which tested dosages ranging from 0 (placebo) to 30 mg. The resulting data, shown in the left-most diagram, indicate a close fit to a Regression line which shows that increasing the dosage (*x* Variable) increases the subject's health outcome (*y* Variable) proportionally.

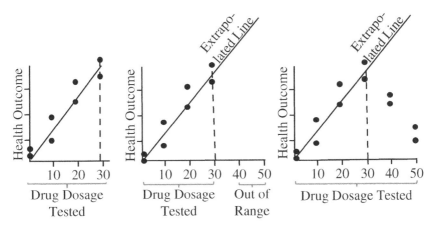

If the experimenters knew little about statistics (and had little common sense), they might be tempted to call a press conference announcing that they had proved that larger and larger doses of the drug produce better and better results (middle graph). They basically would then be extrapolating their findings beyond the range of the dosages (*x* Variable) tested.

Fortunately, wiser heads prevailed, and they announced that their test results were only valid for 0 – 30 mg. Subsequent research tested doses from 40 to 50 mg (right diagram). This showed that doses of 40 and 50 mg actually resulted in poorer health outcomes than those for 40 mg.

Here's a little cartoon that may help reinforce the dangers of extrapolation.

Our experiments show that 3 pills produce results which are 3 times as good as 1 pill.

So, 10 pills should be 10 times as good.

Actually, 10 pills would send you to the hospital.

Related Articles in This Book: *Design of Experiments – Parts 1–3*; *Correlation Part 2*; *Regression Parts 1, 2, 4, and 5*; *Residuals*

REGRESSION – PART 4 (OF 5): MULTIPLE LINEAR

Prerequisite articles: Correlation, Part 2; Regression Parts 1, 2, and 3; and Residuals.

Summary of Keys to Understanding

1. **The same basic principles that apply to Simple Linear Regression and to Residuals also apply to** Regression using two or more variables, **Multiple Linear Regression.**

$$y = b_1x_1 + b_2x_2 + \cdots + b_kx_k + a$$

However**, additional considerations** (and complications) **apply.**

2. **Start by identifying possible Predictor Variables** (the x's), **using** <u>subject matter knowledge</u>. Then <u>collect the data</u>. Then, <u>plot the data</u> **and perform** <u>Correlation Analyses</u> **to identify candidates for the Regression Model.**

3. **The more is NOT the merrier. Drop** x **Variables which** <u>are</u> <u>not</u> **Correlated with the** y **Variable or which** <u>are</u> **Correlated with other** x **Variables.**

4. From a data set of y and corresponding x variables, **statistical software can provide the Regression Model and a number of Statistics for evaluating it.**

$$y = -38.824 + 83.725 \times \text{Bedrooms} + 76.078 \times \text{Bathrooms}$$

5. **A Model cannot be validated with the data used to produce it.** Use Design of Experiments to **design and conduct controlled experiments on the predictions** from the Model.

Explanation

> **1. The same basic principles that apply to Simple Linear Regression and to Residuals also apply to** Regression using two or more variables, **Multiple Linear Regression.**
>
> $$y = b_1 x_1 + b_2 x_2 + \cdots + b_k x_k + a$$
>
> However, **additional considerations** (and complications) **apply.**

In Simple Linear Regression, y is a function of a single x Variable, $y = f(x)$. In Multiple Linear Regression there are more x Variables. $y = f(x_1, x_2, \ldots, x_k)$.

In Simple Linear Regression, the equation for the Regression Model is $y = bx + a$. In Multiple Linear Regression, the following is the equation for the Model. k is the number of Predictor Variables (x's).

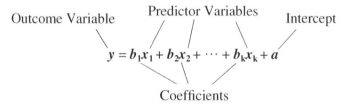

The addition of additional x's is non-trivial. A number of considerations and complications arise in Multiple Regression that do not exist in Simple Regression. This makes it much more complicated:

- There are **several methods and measures for selecting the x variables** to include in the Regression Model.
- The Coefficient of Determination, R^2, **tends to grow solely as a result of increasing the number of x Variables.** However, some other Statistics don't have this shortcoming.
- **You can't plot a 2-dimensional graph with two or more x Variables.** With two x Variables plus one y Variable, the Regression produces a Best Fit 2-dimensional Plane in a 3-dimensional space. Beyond that, we get into an imaginary realm of 4+ dimensions. However, we can and do use plots of y vs. each individual x.
- Usually, there are a number of candidates for inclusion as x Variables in the Regression Model. Various methods and Statistics are employed to winnow these down to a **Best Subset or Subsets.**
- **ANOVA** is often used to help evaluate the Model.

- There may be **Interactions** between or among the *x* Variables. These are normally analyzed during the Designed Experiments which are required for validating the predictive ability of the Model.

> **2. Start by identifying possible Predictor Variables** (the *x*'s), **using subject matter knowledge**. Then **collect the data**. Then, **plot the data and perform Correlation Analyses to identify candidates for the Regression Model.**

Step 1: The first thing to be done does not involve statistics. Use your subject matter knowledge – and that of other subject matter experts, if available – to **identify all measurable causes (*x* Variables) of Variation in the *y* Variable.** It is important to get this right, because once the data are collected, you can't go back and get data on additional *x*'s.

Step 2: Collect the data, **For each subject, you must collect all the *x*-values at the same time.** You can't record someone's blood pressure today and then come back a couple of weeks later to get their weight. This is why Step 1 is so important.

Example
We are developing a Model for predicting House Price (the *y*, in thousands of dollars). Subject matter experts have identified four possible *x*'s (actually, there are more, but let's keep this example simple): House Size, in thousands of square feet; Lot Size, also in thousands of square feet; number of Bedrooms; and number of Bathrooms. We collected this data:

Price	House Size	Lot Size	Bedrooms	Bathrooms
200	2.0	7.5	2	1
300	1.3	5.0	2	2
300	1.8	5.0	3	1
350	1.9	8.0	3	2
340	2.2	8.0	3	2
280	2.5	10.0	2	2
500	3.0	8.0	4	2.5

> **3. The more is NOT the merrier. Drop *x* Variables which are not Correlated with the *y* Variable or which are Correlated with other *x* Variables.**

Simpler – fewer *x* Variables – **is better.** Including a large number of possible *x*'s can be counterproductive, due to interactions among the *x* Variables, increased expense, and complexity of subsequent experiments.

Step 3: The *Part 2* article said, "… plot the data. Then perform a Correlation analysis. Only perform a … Regression if there is a moderately strong Correlation between the Variables." This applies to Multiple Regression as well.

Do a Scatterplot of *y* vs. each candidate *x*. If the Scatterplot shows a roughly linear Correlation, proceed to Step 4; otherwise drop that *x* Variable from consideration.

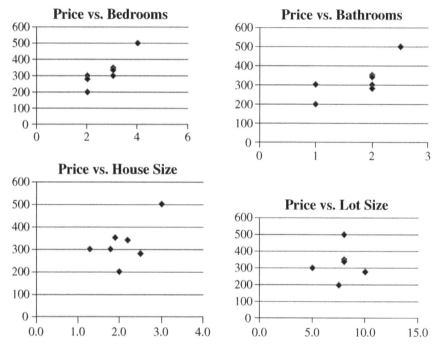

Price vs. Lot Size looks random, so we drop it. As the article *Correlation Part 2* instructs, we don't even bother to calculate the Correlation Coefficient if the plot doesn't look at least roughly linear.

The plots of the other three *x*'s do. So, we can proceed with them to the next step.

Step 4: **Calculate *r*'s, the Correlation Coefficients, for each *x* Variable with the *y* Variable.** Guideline: If $r \geq 0.50$ or $r \leq -0.50$, then there is at least a moderately strong Correlation between that *x* Variable and the *y* Variable. **Drop any *x* Variable which does not have at least a moderately strong Correlation with the *y* Variable.**

After dropping Lot Size as a result of the Scatterplot, we see the remaining three x Variables all have $r > 0.50$. So we retain all three – for now.

While we <u>do want</u> each x to be linearly Correlated with y, we <u>don't want</u>

- **any two x's Linearly Correlated with each other.**
- **any single x Linearly Correlated with all the rest.**

If any two (or more) x Variables are Correlated with each other, that is called Multicollinearity. Multicollinear x Variables would have a similar effect on the y Variable, and the **Regression software would have no good way to determine how to allocate this effect among the two (or more) Multicollinear x Variables**. So, inaccurate Regression Models could result.

Step 5: Measure the Correlation between all pairs of x Variables. **If any pair of x Variables is strongly Correlated, drop one of them.** Drop the one that is less correlated with the y Variable. As one might expect, different experts use different numbers as the clip level for "Strongly Correlated" 0.6 or 0.7 seem to be common. We calculate:

for House Size vs. Bedrooms: $r = 0.57$
for House Size vs. Bathrooms: $r = 0.46$
for Bedrooms vs. Bathrooms: $r = 0.42$
So, we'll keep all three x Variables at this point

> **4.** From a data set of y and corresponding x variables, **statistical software can provide the Regression Model and a number of Statistics for evaluating it.**

Step 6: **Using statistical software, run a Regression Analyses on the remaining x Variables.**

In our example, we obtain this equation for a Regression Model:

$y = -34.75 - (5.349 \times \textbf{House Size}) + (85.506 \times \textbf{Bedrooms}) + (77.486 \times \textbf{Bathrooms})$

-34.75 is the Intercept. The three **Coefficients** corresponding to the three x Variables are -5.439, 85.506, and 77.486.

In addition to this equation for the Regression Model, two other sections of the Regression analysis output – Regression Statistics and ANOVA – describe the Model as a whole. (Another section describes the individual x Variables.)

"R Square," R^2, tells us that 97% of the Variation in the value of the House Price is explained by the Model. Adjusted R Square is almost as good. So this Model is a very good fit for the data. The Standard Error is the Standard Deviation of the Errors (Residuals).

Regression Statistics	
Multiple R	0.985
R Square	0.970
Adjusted R Square	0.955
Standard Error	19.529
Observations	7

The ANOVA section evaluates the Null Hypothesis that the Regression Model explains <u>none</u> of the Variation in the data.

ANOVA	df	SS	MS	F	p-value
Regression	2	48845.938	24422.969	64.040	0.001
Residual	4	1525.490	383.371		
Total	6	50371.429			

The p-value of 0.001 is the Probability that the results obtained are a False Positive, as opposed to the Model being valid. (With some software, p-value is labeled "Significance F.") Here, it is well below the usual clip level for Alpha of 0.05, so this is further corroboration of the validity of the Model.

The following section looks at properties of the individual components of the Model. It will tell us which of the three x Variables we selected make a Statistically Significant contribution to the Model. Those will be included in the final version of the Model.

	Coefficients	Std Error	t-Stat	p-value	Lower 95%	Upper 95%
Intercept	−34.750	40.910	−0.849	0.458	−164.944	95.445
House Size	−5.439	21.454	−0.254	0.816	−73.716	62.838
Bedrooms	85.506	15.002	5.700	0.011	37.763	133.249
Bathrooms	77.486	18.526	4.183	0.025	18.529	136.443

The **Coefficients** for the x Variables (House Size, Bedrooms, and Bathrooms in this example) are the b's in the equation: $y = a + b_1x_1 + b_2x_2 + b_3x_3$. The Coefficient of the Intercept (a) is its value, since it is a constant.

p-value: In this example, we selected $\alpha = 5\%$. If the p-value $\leq \alpha$, then we can conclude that the Coefficient of that x-Variable is different from zero to a Statistically Significant degree, and thus it can be included in the Model.

We see that the p-values for Bedrooms and Bathrooms are comfortably below $\alpha = 0.05$. This tells us that they <u>do</u> make a Statistically Significant contribution to the Model. So we will retain them going forward. However, for House size, the p-value is very large.

<u>Step 7</u>: **The statistical software identifies a best subset or best subsets of x Variables.**

This can get very complicated – testing all possible combinations of subsets. Also, some "common sense" methods of adding or subtracting individual x Variables are less than effective. So it's best to find a statistical software which can do this for you.

For example, the very high p-value for House Size makes it a candidate to be dropped. As we see in the summary below, doing so does not decrease the accuracy of the Model.

# of x Variables	x Variables	R^2
3	House Size, Bedrooms, Bathrooms	0.970
2	Bedrooms and Bathrooms	0.970
2	House Size and Bathrooms	0.649
2	House Size and Bathrooms	0.797

There are a number of methods for selecting best subsets. The best approach may be to select a set or sets of x Variables which optimize one of the Statistics: R^2, Adjusted R^2, or Mallow's C_p:

- **R Square, R^2,** is the square of Multiple R in Multiple Regression. As discussed in Part 2, R^2 is the proportion of the Variation in y that is explained by the Model. Being a Proportion, R^2 ranges from 0 to 1. The larger the value of R^2, the better. The drawback to R^2 is that it tends to gets larger simply as a result of more x Variables being added. (This is another reason why the more is not the merrier.)

$$R^2 = \frac{\text{SSR}}{\text{SST}} = 1 - \frac{\text{SSE}}{\text{SST}} \ (see\ the\ part\ 2\ article)$$

Adjusted R^2, denoted $\bar{R}^2$, adjusts for this shortcoming in R^2. If n is the Sample Size and p is the number of Predictor (x) Variables then,

$$\bar{R}^2 = R^2 - (1 - R^2)\frac{p}{n - p - 1}$$

There are mixed opinions regarding whether it is a substantial improvement.

- **Mallows C_p** makes a similar adjustment. It is preferred by some experts, but it is not often included in the standard output of software and spreadsheets.

For the best subset Bedrooms and Bathrooms Model above, the Regression Model is:

$y = -38.824 + (83.725 \times$ Bedrooms$) + (76.078 \times$ Bathrooms$)$

Step 8: Finally, we need to **analyze the Residuals**, as described in the article *Residuals*.

Remember: Don't extrapolate! In this example Regression analysis, the data ranges of our remaining two *x* variables were: Bedrooms; 2–4 and Bathrooms 1–2.5. So, the Model is only good for houses with Bedrooms and Bathrooms within those ranges.

> **5. A Model cannot be validated with the data used to produce it.** Use Design of Experiments to **design and conduct controlled experiments on the predictions** from the Model.

What we just did was to test the Model with the data that produced it. That was useful in determining the best subset of *x* variables to include in the Model, and to give us confidence that we are on the right track. But we're not done yet. **The whole purpose of Regression is to make predictions about data we have not yet collected.**

So, we must conduct experiments in which, for specific values (levels) of each of our *x* Variables, we predict a *y*-value. There is a discipline called Design of Experiments which describes how to do this in order to get Statistically Significant results. This needs to be done before we can make any definitive conclusions about Cause and Effect between the *x* and *y* Variables.

Related Articles in This Book: *Regression, Parts 1–3 and 5; Correlation Part 2; r, Multiple R, r², R², R Square, R² Adjusted; Alpha, α; Design of Experiments (DOE) – Parts 1–3; p-Value (p); t-tests – Parts 1 and 2; F; ANOVA – Parts 1–4; Residuals; Standard Error*

REGRESSION – PART 5 (OF 5): SIMPLE NONLINEAR

Prerequisite articles: Correlation, Part 2; Regression Parts 1, 2, and 3; and Residuals.

Summary of Keys to Understanding

1. Simple Nonlinear Regression <u>fits a curve</u> to nonlinear x–y data. The y is a function of a single x Variable: $y = f(x)$.

Depending on the shape of the data, there are many different types of curves, resulting from many different kinds of functions.

2. Data shaped like Exponential, Logarithmic, and some other functions can be transformed so that Simple Linear Regression can be used on them.

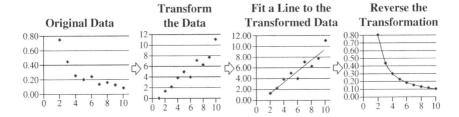

| Original Data | Transform the Data | Fit a Line to the Transformed Data | Reverse the Transformation |

3. If the data curve changes direction, a Polynomial curve can be fit.

4. The usual Regression cautions and restrictions apply:
- **Analyze the Residuals**
- **Don't Extrapolate**
- **A Regression Model cannot be validated with the data used to produce it.** Use Design of Experiments (DOE) to design and conduct controlled experiments on the predictions from the Model.

Explanation

> **1. Simple Non-linear Regression fits a curve to non-linear x–y data. The y is a function of a single x Variable: $y = f(x)$.**
>
> Depending on the shape of the data, there are many different types of curves, resulting from many different kinds of functions.

"**Simple" means there is only one x-Variable** – the same as in Simple Linear. In both Simple Linear and Simple Non-linear Regression, $y = f(x)$. "Simple" is the opposite of "Multiple." In "Multiple" Regression (Linear or Non-linear), there are two or more x Variables: $y = f(x_1, x_2, \ldots, x_k)$. Multiple Linear Regression was covered in Part 4

"**Non-linear" means we try to fit a curve to the data**, as opposed to a line. Multiple Non-linear Regression can be very complicated, and it is beyond the scope of this book.

Even for Simple Nonlinear Regression, the calculations can be involved, and it is advisable to use statistical software which automates them, as opposed to trying to do them with spreadsheets alone. This article will illustrate the concepts, without attempting to show the individual calculations.

"Curve" is a very inclusive term. Depending on the shape of the data, there are many different types of curves, resulting from many different kinds of functions, which can be used. The following are the most common:

Exponential	Logarithmic	Power	Polynomial
$y = a * b^x$	$y = a + (b * \log(x))$	$y = a * x^b$	$y = a_i x^i + a_{i-1} x^{i-1}$ $+ \cdots + a_1 x + a_0$

Exponential and Logarithmic have rapid accelerations or decelerations in the Slope. Power curves have a more gradual change. Polynomial functions can be used for more complex curves, as we'll see later.

> **2. Data shaped like Exponential, Logarithmic, and some other functions can be Transformed so that Simple Linear Regression can be used on them.**

It is much more difficult to fit a curve to data than to fit a line. So, Nonlinear Regression makes use of mathematical Transformations in order to use Simple Linear Regression techniques on Nonlinear data.

Step 1: **Select a type of curve** **that approximates the shape of the data**. The curve type may be Exponential, Logarithmic, Power, Polynomial, or another type. In the example below, the data do not follow a linear pattern. The curve for $y = 1/x$ appears to be a fair approximation. $y = 1/x$ can be rewritten as $y = x^{-1}$, which makes it a Power function.

Original data

x	2	3	4	5	6	7	8	9	10
y	0.75	0.45	0.26	0.20	0.25	0.14	0.16	0.13	0.09

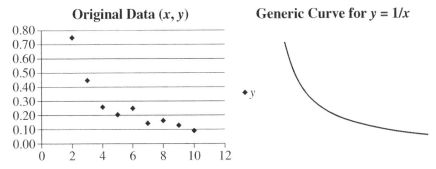

Step 2: **Transform the data:** **For each value of** x**, calculate a Transformed value for** y**, using the inverse of the function for the selected curve**. The specific equation can be determined by solving for x algebraically.

Exponential and Logarithmic functions are, by definition, inverses of each other. That is, if $x = \log_b(y)$, then $y = b^x$.

In our example, $y = 1/x$, so, algebraically, $xy = 1$, and then $x = 1/y$.

So, we apply the Transformation $y = 1/y$, i.e., a Transformed $y = 1/$ the Original y.

Transformed data

x	2	3	4	5	6	7	8	9	10
Transformed $y = 1/$Original y	1.33	2.22	3.85	5.00	4.00	7.14	6.25	7.69	11.11

If the original data follow the shape of the selected curve, the Transformed data – in pairs of values (x, Transformed y) – **should approximately follow a straight line.** And it does, as shown in the plot below left:

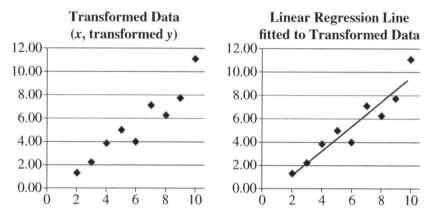

Step 3: Use Simple Linear Regression to fit a line to the Transformed data.

The Simple Linear Regression analysis comes up with the following equation for the fitted line:

$$y = -0.847 + 1.041x$$

It is pictured above in the plot above right. The analysis also tells us that R Square = 0.893, Adjusted R Square = 0.878, and the p-value is 0.0001. So this is a good fit of a straight line to the Transformed data.

Step 4: **Reverse Transform the fitted Regression Line to produce the equation for the fitted Regression Curve.**

To recap: Curve selected in Step 1: $y = 1/x$

Transformed $y = 1/$ Original y

So, to reverse this transformation

y for Fitted Regression Curve = $1/y$ from Fitted Regression Line

In our example, $y = 1/(-0.847 + 1.041)$.

If we plug in x's from 2 to 10, we get:

x	2	3	4	5	6	7	8	9	10
Y from fitted Curve	0.810	0.439	0.301	0.229	0.185	0.155	0.134	0.117	0.105
Original y (for comparison)	0.75	0.45	0.26	0.20	0.25	0.14	0.16	0.13	0.09

The y's from the fitted Curve are those predicted by the Simple Nonlinear Regression Model, $y = 1/(-0.847 + 1.041)$.

One might have been originally skeptical about the distortions caused by Transforming and Reverse Transforming. But as we can see from the Scatterplot below, good results can be obtained.

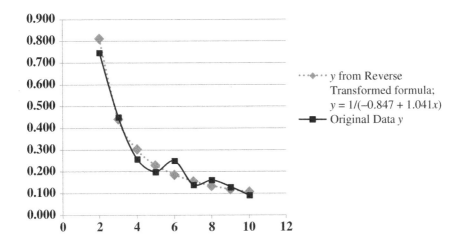

> **3. If the data curve changes direction, a Polynomial curve can be fit.**

A Polynomial has an equation of the form:

$$y = b_k x^k + b_{k-1} x^{k-1} + \cdots + b_1 x + a$$

Note that there is just one x Variable, but it is raised to various powers, starting with the power of 2: x^2. (If there were only a power of 1, the equation would be that of a straight line.) The b's are Coefficients and the a is an Intercept.

A "second degree," also known as "second order" or "Quadratic," Polynomial is of the form:

$$y = b_2 x^2 b_1 x + a$$

For example, $y = 3x^2 + 7x + 11$. A second order Polynomial has one change in direction. As x increases, y increases and then decreases (or y decreases and then increases). Two examples are pictured above. These shapes are Parabolas.

A "third degree," aka "third order" aka Cubic" Polynomial has an x^3 term and changes direction twice.

A kth degree Polynomial has $k-1$ changes in direction.

Simpler is better. It is usually not necessary to go beyond three orders. Larger orders are harder to work with. Also, they may be too closely associated with the idiosyncrasies of the data provided in a particular Sample, and they may not be generally applicable to data in other Samples from the same Population or Process.

> **4. The usual Regression Cautions and Restrictions apply:**
> - **Analyze the Residuals**
> - **Don't extrapolate**
> - **A Regression Model cannot be validated with the data used to produce it.** Use Design of Experiments (DOE) to **design and conduct controlled experiments on the predictions** from the Model.

These are covered in the articles,

- *Residuals*
- *Regression – Part 3, Analysis Basics*
- *Design of Experiments*

See also: Correlation – Part 2; Regression – Part 3: Simple Linear

REJECT THE NULL HYPOTHESIS

Summary of Keys to Understanding

 1. **"Reject the Null Hypothesis" is one of two possible conclusions from a Hypothesis Test.**
The other is "Fail to Reject the Null Hypothesis"

 2. **The Null Hypothesis (symbol H_0) states that that there is no**
- **difference, or**
- **change, or**
- **effect**

 3. **So, to Reject the Null Hypothesis is to conclude that there is a**
- **difference, or**
- **change, or**
- **effect**

A statistician responds to a marriage proposal:

Will you marry me?

I Reject the Null Hypothesis.

 Yes! "Reject" means "Yes", because the Null Hypothesis is a negative, and rejecting a negative results in a positive.

Explanation

 1. **"Reject the Null Hypothesis" is one of two possible conclusions from a Hypothesis Test.**

The other is "Fail to Reject the Null Hypothesis"

As stated in the article in this book, "*Hypothesis Testing – Part 1: Overview*," Hypothesis Testing is one method of Inferential Statistics. It is a method for answering questions about a Population or a Process by analyzing data from a Sample.

The *Hypothesis Testing Part 2* article describes the five steps in this method. In Step 2, we select a value for the Level of Significance, Alpha (α). In Step 4, the analysis calculates a value for p, the Probability of an Alpha Error.

Step 5 is to come to a conclusion about the Null Hypothesis by comparing p to Alpha. There are only two possible conclusions:

- If $p \leq \alpha$, **Reject the Null Hypothesis** (the subject of this article), or
- **Otherwise, Fail to Reject the Null Hypothesis** (the subject of a separate article)

 2. **The Null Hypothesis** (symbol H_0) states that that **there is no**
- **difference, or**
- **change, or**
- **effect**

There is also an article devoted to the Null Hypothesis with a number of examples. However, a picture may be worth a thousand words.

Will you marry me?

You might think that these cartoon examples are kind of silly and are unbecoming of a serious tome about a precisely defined science. But since this is not that serious of a tome, and since statistics is not always a precisely defined science, maybe it's not so unbecoming.

Also, silly can be beneficial in this case, because the silliness may be what helps us to remember that **rejecting is a positive result** because **the Null Hypothesis is a negative statement.**

> **3. So, to Reject the Null Hypothesis is to conclude that there is a**
> – **difference, or**
> – **change, or**
> – **effect**

In our illustration,

– If the woman statistician were to agree to get married, there <u>would be</u> a change in their relationship; there <u>would be</u> something different.
– So by Rejecting the Null Hypothesis, she has agreed that there has been a change – they are now engaged to be married.

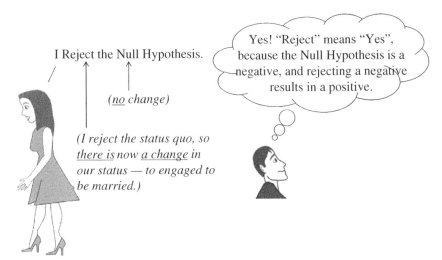

I Reject the Null Hypothesis.

(no change)

(I reject the status quo, so <u>there is</u> now <u>a change</u> in our status — to engaged to be married.)

Yes! "Reject" means "Yes", because the Null Hypothesis is a negative, and rejecting a negative results in a positive.

What about the Alternative Hypothesis?

As explained in the article by that name, experts have conflicting views whether or not an Alternative Hypothesis is necessary and how it should be defined. In any event,

• <u>if</u> an Alternate Hypotheses (H_A) was stated at the same time as the Null Hypothesis (H_0), and

- if H_A is defined as the hypothesis which must be true if H_0 is false (i.e., the two are collectively exhaustive),

then
to Reject the Null Hypothesis is to accept the Alternative Hypothesis.

The "Leave Well-Enough Alone" caution:

If you're comfortable that things are clear now, and that you understand the concept, feel free to stop here. But we can also continue with another explanation which may help in understanding the concept.

- The Null Hypothesis is verbally stated as a Negative. In this situation, it would be,
  "There is no change in the relationship."
  (If it helps, think of a negative statement as having a value of −1.)
- Rejecting adds a "not" making it a double negative, which adds to the confusion.
  "There is not no change in the relationship"
  (There are now two −1s.)
- A double negative is logically equivalent to a positive.
  "There is a change in the relationship."
  $(-1 \times -1 = 1)$
- **So Rejecting the Null Hypothesis is the same as answering Yes to the Yes/No Question**:
  "Is there a change in the relationship?" Answer: Yes
  The change is from not being engaged to being engaged.

Related Articles in This Book: *Null Hypothesis*; *Fail to Reject the Null Hypothesis*; *Alternative Hypothesis*; *Hypothesis Testing – Part 1: Overview*; *Hypothesis Testing – Part 2: How To*; *Alpha, p-Value, Critical Value, and Test Statistic – How They Work Together*

RESIDUALS

Summary of Keys to Understanding

1. Residuals are an important part of Regression analysis. **The value of the Residual of any (x, y) data point is the value of the y predicted by the Regression Model** (denoted by y with a "hat") **minus the actual value of y** observed in the data point.

$$\text{Residual} = \text{predicted} - \text{actual} = \hat{y} - y$$

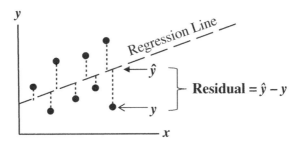

2. Residuals represent the **Error** – in the Regression Model – the **Variation** of the Outcome Variable y **which is unexplained by the Model**. They **must be analyzed to ensure that this Variation is truly unexplainable** by any other factors (x Variables) which are not included in the Regression Model.

3. Standardize the Residuals before analyzing them.

4. Residuals must:
- be **Random**
- be **Normally Distributed**
- **not be Correlated with any x Variable**
- **not be Autocorrelated** (relative to time sequence)
- have **Constant Variance**
- have **no unexplained Outliers**

Explanation

> **1.** Residuals are an important part of Regression analysis. **The value of the Residual of any (x, y) data point is the value of the y predicted by the Regression Model** (denoted by y with a "hat") **minus the actual value of y** observed in the data point. **Residual $= \hat{y} - y$.**

Virtually no Regression Model is a perfect fit for real-world data. For any given value of x (or multiple x's) the predicted (aka "expected") value of y calculated by the Model will usually <u>not</u> be identical to the actual value of y for that x (or x's) in the data. (Assuming there is a data point for that particular value of x). This difference is considered an Error in the Model. It is called a "Residual."

"Residue," according to Dictionary.com means "*something that remains after a part is removed, disposed of, or used; remainder; rest; remnant.*" **After we remove the Variation in Y which is explained by the Regression Model, what's left** (the residue) **is called a Residual.** The remaining Variation is – for each (x, y) data point – the distance from that point to the Regression Line, as measured along the y-axis.

$$\text{Residual} = \text{predicted} - \text{actual} = \hat{y} - y$$

In the diagram below, the Regression Model is graphed as the diagonal dashed line. "y" is the actual value of y for the data point (x, y) and the y-hat is the predicted value of y which is on the Regression Line for that value of x.

The values of the Residuals are the lengths of the vertical dotted lines. Residuals for points above the Regression Line are negative.

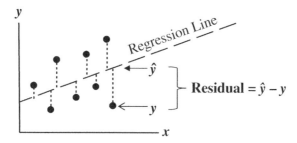

The value of the Residual of any (x, y) data point is the value of y Predicted by the Regression Model (denoted by y with a "hat") **minus the Actual value of y.**

2. Residuals represent the **Error** – in the Regression Model – the **Variation** of the Outcome Variable y **which is unexplained by the Model**. They **must be analyzed to ensure that this Variation is truly unexplainable** by any other factors (x Variables) which are not included in the Regression Model.

In the article *Regression – Part 1: Sum of Squares*, we learned **that the Sum of Squares Total (SST) is a measure of the total Variation in y**. Regression attempts to account for this Variation as being caused by one or more x Variables:

$$y = f(x) \quad \text{or} \quad y = f(x_1, x_2, \ldots, x_n)$$

This total Variation, SST, can be divided into two parts: **the Variation explained by the Regression Model,** which is the Sum of Squares Regression (SSR), **and the remaining unexplained Variation,** which is the Sum of Squares Error, SSE. SSE is the Variation **exhibited in the Residuals.**

$$\textbf{SST} = \textbf{SSR} + \textbf{SSE}$$

The *Regression – Part 2* article, defines R^2, a measure of the Goodness of Fit of the Model.

$$R^2 = 1 - \frac{\text{SSE}}{\text{SST}}$$

But, **it's not just the size of SSE that is important.** The article *Regression – Part 3, Analysis Basics* goes on to say that having a good fit, – a high enough value for R^2 – is not good enough. **It's the kind of errors (Residuals) comprising SSE that is also important.** But first, …

 3. Standardize the Residuals before analyzing them.

Residuals are in the real-world units of the y-values in the data, e.g., kilograms, dollars, seconds. Residual analysis (aka Residual diagnostics) will be easier to understand if we convert them to **Standardized Residuals**. For any individual residual,

$$\text{a Standardized Residual} = \frac{\textbf{the Residual}}{\textbf{Standard Deviation of all Residuals}}$$

So, Standardized Residuals are in units of Standard Deviations. As we'll see below, this will help us make better use of the Normal Distribution in our analysis.

> **4. Residuals must:**
> - be **Random**
> - be **Normally Distributed**
> - **not be Correlated with any** x **Variable**
> - **not be Autocorrelated** (relative to time sequence)
> - have **Constant Variance**
> - have **no unexplained Outliers**

If the Regression Model has accounted for all sources of Variation (the x's), then any remaining Variation in the value of the y Variable – **the Residuals – must be just Random noise.**

So, a Scatterplot of Residuals against y-values should illustrate Randomness.

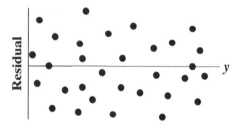

There should be no pattern. A pattern forming a curve may indicate that Nonlinear Regression is needed. If there is a pattern approximating a slanted line, then there may be an additional x Variable that needs to be added to the Model. Other patterns are shown on the following page.

Random noise should be Normally Distributed. This can be illustrated with a Histogram of the Residual Values and statistically verified with a test for Normality, such as the Anderson–Darling test.

Using Standardized Residuals with a Normal Distribution enables us to use the Empirical Rule (aka the "68 – 95 – 97 Rule"), as shown below

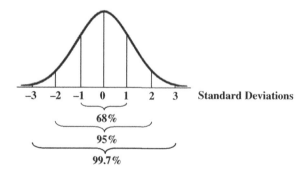

Residuals must <u>not</u> be Correlated with any *x* Variable. Check this with a Scatter Plot of the Residuals against each of the *x* Variables. **There should be no pattern.** Or the pattern should approximate a horizontal band, which indicates that the Residuals do not vary as the *x* varies. Also, evaluate the Coefficients of Correlation, the *r*'s.

Residuals must <u>not</u> be Correlated with each other (Autocorrelation). This can be seen in a time sequence plot with the time period as the horizontal axis and the value of the Residual on the vertical axis. **There should be no pattern.** Below is an example what we should <u>not</u> be seeing.

Not Random:
<u>Autocorrelation:</u>

Residuals must have Constant Variance. The Spread of the Residuals should <u>not</u> increase or decrease over time or in concert with an increase in an *x* Variable. We should <u>not</u> see a "megaphone" shape on a scatter plot.

Not Random:
Megaphone shape, <u>**Lack of**</u>
<u>**Constant Variance:**</u>

Outliers are a potential cause for concern and should be investigated. Definitions vary regarding what exactly is an Outlier. Any point beyond about 2.5 Standard Deviations from the Mean has about a 1% chance of occurring in a Standard Normal Distribution (which is the Distribution that Standardized Residuals should follow if they are random). So, that is a reasonable definition for these purposes.

A Control Chart or a "Box-and-Whiskers" plot can be used to identify Outliers. (With the latter, an Outlier will be outside a "whisker.") With small Sample Sizes, Outliers can have a disproportionate effect on the calculations.

In addition, **Outliers are considered to be** outside the Distribution and **due to "Special Causes" which are outside of the Process or Population which generated the data**. So, Outliers need to be investigated to see what may be causing them.

Related Articles in This Book: *Regression, Parts 1–5*; *Standard Deviation*; *Normal Distribution*; *Correlation Part 2*; *Variance*; *Control Charts – Parts 1 and 2*; *Charts, Graphs, Plots – Which to Use When*

SAMPLE, SAMPLING

Summary of Keys to Understanding

1. To get a good Sample which accurately represents the Population or Process from which it was collected, we need to minimize Sampling Error and Sampling Bias.

   Sampling Error can generally be decreased by increasing the Sample Size.

2. Sampling Bias causes a Sample to <u>not</u> be representative of the Population or Process.

3. If possible, obtain a Sampling Frame which includes all members of the Population or Process.

4. Randomness is likely to be representative. Simple Random Sampling (SRS) can be the most effective method of achieving Randomness.

5. Other Sampling methods include Systematic, Stratified, and Clustered.

6. Statistical Process Control uses a special type of Sample – the Rational Subgroup – to Block the effects of potential causes of Variation.

Statistics from A to Z: Confusing Concepts Clarified, First Edition. Andrew A. Jawlik.
© 2016 John Wiley & Sons, Inc. Published 2016 by John Wiley & Sons, Inc.

Explanation

> 1. **To get a good Sample which accurately represents the Population or Process from which it was collected, we need to minimize Sampling Error and Sampling Bias.**
>   **Sampling Error can generally be decreased by increasing the Sample Size.**

In <u>Descriptive</u> Statistics, we have all the data on an entire Population or Process. So, we can calculate numerical values which <u>describe</u> its statistical properties. For a Population or Process, these are properties called Parameters, e.g., the Mean or Standard Deviation.

But most often, we don't have all the data from the entire universe under consideration. So, we collect a Sample of data, and we calculate a statistical property for the Sample. This property of a Sample is called a Statistic. For every Parameter, there is a corresponding Statistic. **In <u>Inferential</u> Statistics, the value of the Sample Statistic becomes our estimate** (<u>inference</u>) **of the value of its corresponding Population or Process Parameter.**

Even if we did everything right in collecting our Sample, there is a calculable Probability that our particular Sample is not a good representative of the Population or Process. For example, we know that the Population of coin flips is 50% heads and 50% tails. We also know that many Samples of coin flips will not be 50/50; in fact some Samples will be more like 60/40.

This is called the Sampling Error. It is not an error in the sense that a mistake has been made. It is just acknowledging the fact that <u>some</u> of the Samples we can potentially collect are <u>not</u> going to be representative of the Population or Process. (See the article *Alpha and Beta Errors*.)

We can minimize Sampling Error to an acceptable level by having a sufficiently large Sample Size. This subject is addressed in the two articles in this book on Sample Size.

But no increase in Sample Size will address Sampling Bias.

> 2. **Sampling Bias causes a Sample to <u>not</u> be representative of the Population or Process.**

Sampling Bias does not occur by chance, as does Sampling Error. It occurs because a non-representative method of collecting data was used. Let's say we are conducting a poll to determine how voters plan to vote in a 2-candidate race. We get our clipboard and stand outside a women's clothing store in a mall one evening and try to question every person that happens by. There are number of sources of **Selection Bias** in this approach which can lead to a Sample that is not representative of the Population:

- More women than men will be going in and out of a women's clothing store.
- The men would more likely be married or have a girlfriend or a close female relative.
- People who work the evening shift will be excluded.
- People who are not eligible to vote can be included.
- If the mall caters to either high-income or low-income customers, the other income level will be under-represented.
- If the mall is in the city center, then rural and suburban voters will be under-represented.

Response Bias is a type of Sampling Bias. Almost all polls and surveys suffer from Response Bias: many people will not take the time to respond to the questions.

- People with more time on their hands are more likely to respond. This can lead to over-representation of retired people and the unemployed (and people <u>without</u> two screaming toddlers clinging to them).
- People with strong opinions about the question are more likely to respond.
- Patient people are more likely to respond than impatient people.

Sampling other than for polls and surveys can also have Selection bias. For example, in manufacturing, inspectors of physical items tend to select for their Sample those items which have a visible defect.

 3. If possible, obtain a Sampling Frame which includes all members of the Population or Process.

A Sampling Frame is a list which identifies all possible units that can be selected for our Sample. An example would be a list of registered voters, or a numbering scheme for uniquely identifying each item produced in a manufacturing run.

A Sampling Frame can be used in the Sampling methods described below.

 4. Randomness is likely to be representative. Simple Random Sampling (SRS) can be the most effective method of achieving Randomness.

There is no Bias in true Randomness. **In a Random selection** method, **each possible Sample of a given Sample Size is equally likely to be selected**. So, **a Random selection method is Unbiased**.

So, if the Sample Size is large enough to reduce Sampling Error to a tolerable level *(see the articles on Sample Size)*, the Sample will be as representative as possible.

To implement the Simple Random Sampling (SRS) method, a randomizing technique is needed. This can take the form of "picking numbers out of a hat" or using a computer random number generator. (A spreadsheet can provide the latter.) In the latter case, the list of units in the Sampling Frame is numbered, and the units corresponding to the generated numbers become the Sample.

 5. Other Sampling methods include Stratified, Clustered, and Systematic.

There are many different Sampling methods. Here are three of the most common.

Stratified Sampling

- Divide the Population or Process into homogeneous groups (strata).
- Select a Simple Random Sample from each group. The Sample Size for each group corresponds to a known Proportion of the group in the Population or Process.

To use this method, you must know the Proportion. Let's say that our Population is the student body of a college. We know that 55% are women and 45% are men. We define two homogeneous groups – women and men. We want our Sample to have the same gender Proportions as the Population. So, for a Sample of size $n = 100$, we Randomly select 55 women and 45 men.

Advantage: Avoids selecting a Sample which we know is not representative – at least with regard to the Proportions of the homogenous groups.

Disadvantage: Can't be used when there are no homogeneous subgroups.

Clustered Sampling

- Divide the Population or Process into small clusters (e.g., city blocks)
- Select a Simple Random Sample of these clusters
- Collect data from each unit within each cluster

Advantages: It can be less time-consuming and less expensive. For example, the Population is the inhabitants of a city, and a cluster is a city block. We randomly select an SRS of city blocks. There is less time and travel involved in driving to a limited number of city blocks and then walking door to door, compared with traveling to more-widely-separated individuals all over the city. Also, one does not need a Sampling Frame listing all individuals, just all clusters.

Disadvantage: The increased Variability due to between-cluster differences may reduce accuracy.

Systematic Sampling

A Sampling Frame, say a Population list or a sequential manufacturing production run, is required. Number the Sampling Frame from 1 to N.

- Randomly select or generate a number, j, to be the first unit selected for the Sample. (Don't just start with 1, because that would not be Random.)
- Select the kth item after that to be the second item selected.
- Repeat until the desired Sample Size, n, is reached.

k can be randomly selected, but it needs to be small enough to give you the desired Sample Size, n. So, you could calculate it this way: If N is the total number of items in the Population or Process and j is the random number generated for the first item selected then k could be $(N-j)/n$. Round down to an integer.

Example: $N = 300$, $n = 30$, and $j = 6$, so the 6th unit is selected to be the first unit in the Sample.

$k = (300 - 6)/30 = 9.8$, Round down to 9

Select the 6th item, then the 15th, 24th, 33rd, etc.

Advantage: If the Sampling Frame is ordered in some obvious or non-obvious manner, Systematic Sampling avoids a Sample comprised of a disproportionate number of units from the top or bottom of the range.

Disadvantages: If the Sampling Frame has periodic variations which coincide with k, the Sample will not be representative.

> **6. Statistical Process Control uses a special type of Sample – the Rational Subgroup—to Block the effects of potential causes of Variation.**

Up to this point in the article, the Sampling has been for use in Inferential Statistics. In Statistical Process Control (SPC), we need a special kind of Sampling method to determine whether a Process is under Control or not. (See the article *Control Charts – Part 1: General Concepts and Principles*.)

In SPC, we collect a number of small Samples in order to identify Variation over time. These are a special kind of Sample called the Rational Subgroup. The Rational Subgroups (Samples) could be as small as 4 or 5, and **we collect 25 or more of them**. **Rational Subgroups are comprised of individual data points collected from a Process under the same conditions**, for example,

- the same operation
- the same operator
- within a narrow timeframe

These conditions are kept the same, because we want to eliminate (Block) them as potential causes of Variation. We can calculate a Statistic (e.g., the Mean, Proportion, Standard Deviation, or Range) for each Rational Subgroup. **It is the value of that Statistic which gets plotted in the Control Chart.**

Related Articles in This Book: *Inferential Statistics*; *Alpha and Beta Errors*; *Errors – Types, Uses, and Interrelationships*; *Sample Size – Part 1: Proportions/Percentages*; *Sample Size – Part 2: Measurement/Continuous data*; *Control Charts – Part 1: General Concepts and Principles*

SAMPLE SIZE – PART 1 (OF 2): PROPORTIONS FOR COUNT DATA

Summary of Keys to Understanding

 1. **Minimum Sample Sizes are calculated very differently for Count data and Measurement Data.** This article is about **Sample Sizes for Proportions of Count data.**

 2. The report of the results of our statistical analysis might use wording like this:

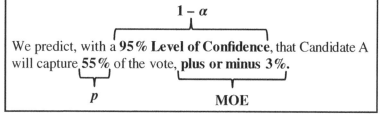

We predict, with a **95% Level of Confidence**, that Candidate A will capture **55%** of the vote, **plus or minus 3%.**

p MOE

 3. **When there is an estimate, $\hat{p}$, for the Population/Process Proportion, the formula for the minimum Sample Size is**

$$n = (\hat{p})(1 - \hat{p})(z_{\alpha/2})^2/(\text{MOE})^2$$

When there is not an estimate, or if you want to take the most conservative approach, set $\hat{p} = 0.5$ and the formula becomes

$$n = (0.25)(z_{\alpha/2})^2/(\text{MOE})^2$$

There are websites that will do these calculations for you.

 4. **The following things increase the minimum Sample Size:**
- **Higher Level of Confidence (i.e., smaller value of Alpha) selected**
- **Smaller Margin of Error specified**
- **Estimated Proportion closer to 0.5**

 5. **After a certain point, larger Sample Sizes yield diminishing returns in accuracy.**

Explanation

 1. Minimum Sample Sizes are calculated very differently for Count data and Measurement Data. This article is about **Sample Sizes for Proportions of Count data.**
The Part 2 article is about Sample Sizes for Measurement/ Continuous data.

A Proportion is a percentage expressed as a decimal. So 50% is 0.50 and 100% is 1.0. Statistical formulas usually use the Proportion format.

Proportions are calculated from Count (aka Discrete) **data.** These are non-negative integer numbers, e.g., 0, 1, 2, 3, etc.

Examples of Proportions of Count Data

Count	Sample Size	Proportion
66 people said they'd vote for Candidate A	120 people were surveyed	$66/120 = 0.55$
8 people preferred strawberry ice cream	20 people in a focus group	$8/20 = 0.40$
6 defective items	production run of 1000	$6/1000 = 0.006$

The symbol for a Proportion is p. That is also the symbol for Probability. The two concepts are related. If the Proportion of people favoring Candidate A is 0.55 then the Probability of any one person favoring Candidate A is 0.55.

If all you want is a quick number – without understanding what's behind it – here are the minimum Sample Sizes for a 95% Confidence Level (the most common) and for several values of the Margin of Error (symbol MOE or E).

95% Confidence Level (the most common)

MOE	1%	2%	3%	4%	5%	6%	7%	8%	9%	10%
Sample Size (n)	9604	2401	1068	601	385	267	196	151	119	97

These results assume you don't know the Population Size (N). If you do, divide the Sample Size above by $1 + n/N$. But if you have to do that, you might as well just do a web search on "Sample Size Calculator" and just enter the relevant numbers on one of those websites.

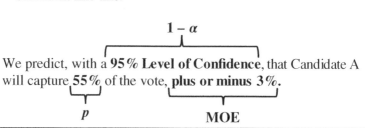

2. The report of the results of our statistical analysis might use a statement like this:

$$1 - \alpha$$

We predict, with a **95 % Level of Confidence**, that Candidate A will capture 55 % of the vote, **plus or minus 3 %.**

p MOE

As we'll see in Key to Understanding #3, the formula for calculating n, the Sample Size, includes four symbols, α, p, MOE, and z. The statement above helps explain what α, p, and MOE (Margin of Error) are about. z is derived from α, as we'll explain later.

Alpha, α can be thought of as the "Lack-of-Confidence" Level

In other contexts, Alpha is called the Level of Significance, and it is involved with the concept of Alpha Error. For our purposes here, it may be more helpful to think of it colloquially as a "Lack-of-Confidence" Level. It is the opposite of the Confidence Level.

In statistics, we don't have a symbol for the Confidence Level, but we do for its opposite. The symbol is the Greek letter Alpha (α).

Level of Confidence $= 1 - \alpha$, so $\alpha = 1 -$ Level of Confidence

(When dealing in percentages, we would use 100% instead of 1.)

The person performing the statistical analysis selects a value for Alpha prior to collecting data for the Sample. **If we want to be 95% Confident that our Sample is representative of the overall Population or Process, that means that we are willing to accept the fact that 5% of the time we will collect a Sample which is not representative.** In that case, we would select $\alpha = 5\%$. (That is the value most commonly selected.)

You may be wondering why you have to settle for 5% or some other number. Why can't you have 0%? Statistically, the only way you can get $\alpha = 0\%$ is if you poll 100% of the Population or Process. However, you could select 1%, for example, but that could negatively affect other things, as we'll see later.

MOE is the Margin of Error, sometimes denoted by "E," instead of MOE.

The actual Population or Process Proportion will very likely <u>not</u> be the exact number which we calculate for p from Sample data. But, we would like it limited to a narrow range. <u>How</u> narrow is specified by the Margin of Error, MOE. MOE is one-half the width of the "plus or minus" range. So, if we say "plus or minus 3%," then MOE = 3%.

We specify the values of Alpha and Margin of Error prior to collecting the Sample data and beginning the analysis. Both of these are involved in calculating how large our Sample will need to be, as can be seen in the following formula.

> **3. When there <u>is</u> an estimate, $\hat{p}$, for the Population/Process Proportion, the formula for the minimum Sample Size is**
>
> $$n = (\hat{p})(1 - \hat{p})(z_{\alpha/2})^2/(\text{MOE})^2$$
>
> **When there <u>is not</u> an estimate, or if you want to take the most conservative approach, set $\hat{p} = 0.5$ and the formula becomes**
>
> $$n = (0.25)(z_{\alpha/2})^2/(\text{MOE})^2$$
>
> There are websites that will do these calculations for you.

This formula assumes you don't know the Population Size (N). If you do know N, divide the n above by $1 + n/N$.

Note: There are websites which will do all these calculations for you. You just have to bring the inputs: your selected values for Alpha (α) and the Margin of Error (MOE), the estimate (or default) Proportion ($\hat{p}$), and the value of N, if known.

$\hat{p}$ is an estimate (or a default) for the actual Proportion of the Population or Process

The formula above was derived from some other equations which assumed that we knew the true Proportion for the Population or Process as a whole. For our purposes, obviously, we don't, or we wouldn't be taking a Sample. So, we must make an estimate or use a conservative default value for $\hat{p}$ (pronounced "p-hat").

The estimate must be from the same Population or Process. It can be another survey, for example, or an earlier sampling of a production run.

With an estimate, there is always a chance for error. Some things may have been different in the other survey, or some Factors may have changed since the earlier production run.

The most conservative approach would be to use the **default value of $\hat{p}$** which gives the maximum value for the product $(\hat{p})(1 - \hat{p})$. That is, $\hat{p} = 0.5$ **(50%)**. As demonstrated in the following table, **the closer p is to 0.5, the larger the value of the product of $\hat{p}$ multiplied by $1 - \hat{p}$, and thus, the larger the value of n.**

$\hat{p}$	$1 - \hat{p}$	$\hat{p}(1 - \hat{p})$
0.10	0.90	0.0900
0.25	0.75	0.1875
0.40	0.60	0.2400
0.50	**0.50**	**0.2500**
0.90	0.10	0.0900

Earlier, we explained what MOE and α are. What is left to explain is z with a subscript of $\alpha/2$.

$z_{\alpha/2}$ is the value of z for a given value of $\alpha/2$

z is a Test Statistic. That means it is calculated from Sample data, and that it has an associated Probability Distribution. z is uniquely suited for our purpose. Other Test Statistics (like t, for example) have a different Distribution for each different value of n, the Sample Size. That wouldn't work for us, because we don't have a number for the Sample Size. We're trying to develop a formula for calculating that number.

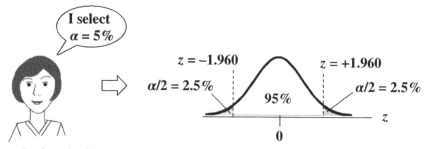

In the Distribution pictured, the height of the curve above any point on the horizontal (z) axis is the Point Probability of the value of z directly below it. The shaded areas represent Cumulative Probabilities of ranges of points. They extend outward to infinity.

We split the 5% Alpha area in half, and we position the resulting two 2.5% areas under the left and right tails of the curve. Since z is a Test Statistic, we can find out from tables or software that the two shaded 2.5% Cumulative Probability areas are bounded by the values of $z = 1.96$ and -1.96. This gives us the value of $z_{\alpha/2} = 1.96$.

So if we select $\alpha = 5\%$, we get $z = 1.960$. If we specify that we want the Margin of Error to be 3%, we can use the default formula to calculate a Sample Size:

$$n = (0.25)(z_{\alpha/2})^2/(\text{MOE})^2$$
$$= (0.25)(1.960)^2/(0.03)^2$$
$$= 1067.11$$

Round up (always) to **1068**

So, we will need to poll at least 1068 people to be 95% Confident with a 3% Margin of Error.

What if we can't afford the time or money to collect data of the calculated Sample Size?

We can sacrifice some accuracy in either the Level of Confidence or the Margin of Error or both.

Let's say we can only afford to poll 625 people. What can we do? With a little algebra, we see that we can plug in a value for n and then calculate either MOE or $z_{\alpha/2}$. Here's how we calculate MOE, given $n = 625$, $\hat{p} = 0.5$, and $z_{\alpha/2}$:

$$n = (0.25)(z_{\alpha/2})^2/\text{MOE}^2$$
$$625 = (0.25)(1.96)^2/\text{MOE}^2$$
$$\text{MOE}^2 = (0.25)(1.96)^2/625$$
$$\text{MOE} = (0.5)(1.96)/25 = 0.0392$$

So, if we reduce the minimum Sample Size from 1068 to 625, that increases the Margin of Error from 3% to about 4%.

What if we were willing to go to a lower Confidence Level (higher level of Alpha) while keeping the MOE at 0.03 and the Sample Size at 625? If we go back to the formula for n, set n to 625 and MOE to 0.03, a little algebra will give us $z_{\alpha/2} = 1.5$.

From tables or software we see that this gives us $\alpha = 0.134$. So, we must be willing to tolerate a 13.4% Probability of an Alpha Error (False Positive), if we want to decrease the Sample Size from 1068 to 625, while keeping a 3% Margin of Error.

For a given Sample Size, Alpha and Margin of Error affect each other inversely.

- If we select a lower value of Alpha (which means a higher Level of Confidence), the Margin of Error increases.

- If we select a higher value for Alpha, the Margin of Error decreases.
- **The only way to reduce both is to increase the Sample Size.**

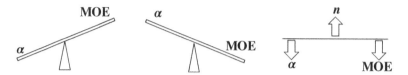

 | **4. The following things increase the minimum Sample Size:**
- **Higher Level of Confidence (i.e., smaller value of Alpha) selected**
- **Smaller Margin of Error specified**
- **Estimated Proportion closer to 0.5**

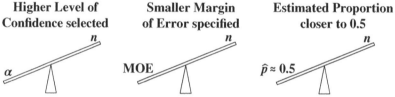

If we keep everything else the same, and we …

- **select a lower value for Alpha, α, (which means a higher Level of Confidence) or**
- **specify a smaller Margin of Error, MOE,**

then the minimum Sample Size will increase.

Also, as we showed earlier, **if we use an estimated Proportion, p $(1 - p)$ is higher – and as a result the Sample Size is larger – when p is nearer 0.5.**

 | **5. After a certain point, larger Sample Sizes yield diminishing returns in accuracy.**

Increasing "accuracy" here means a lower value for Alpha and/or the Margin of Error. We start with the formula,

$$n = (0.25)(z_{\alpha/2})^2 / \text{MOE}^2$$

Note that the two terms describing types of errors, $z_{\alpha/2}$ and MOE, are squared, while n is not. If we solve for either one of them, we're going to get a square root of n in the denominator. For example,

$$\mathbf{MOE} = (0.5)(z_{\alpha/2})/\sqrt{n}$$

So, any reduction in MOE is proportional to the square root of n, not to n itself. For $\alpha = 5\%$, we saw that $n = 1068$ gave us MOE = 3%. If we increase n by about 1000, we can reduce MOE to 2%. If we increase n by another 1000, we'll reduce MOE to only 1.79%. This diminishing returns effect continues to get worse after that.

Related Articles in This Book: *Proportion*; *Sample Size – Part 2: for Measurement/Continuous Data*; *Margin of Error*; *z*; *Test Statistic*

Note, this article addresses a limited aspect of Alpha. There are a number of other articles on Alpha in this book, but they may be unnecessarily confusing if you are just interested in Sample Size right now.

SAMPLE SIZE – PART 2 (OF 2): FOR MEASUREMENT/CONTINUOUS DATA

Summary of Keys to Understanding

 1. **Measurement** (aka Continuous) **data contain more information than Count data, so smaller Sample Sizes can achieve good accuracy.**

 2. **If your focus is on avoiding an Alpha Error (False Positive), you can calculate minimum Sample Size for tests of the Mean using:**
   - α, the selected Level of Significance
   - σ, Standard Deviation of the Population or Process (or an estimate of it)
   - **MOE,** the desired Margin of Error

 3. **All other things being equal, an increase in Sample Size (n) reduces all types of Sampling Errors**, including Alpha and Beta Errors and the Margin of Error.

  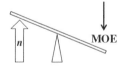

 4. **A larger Sample Size enables smaller differences (Effect Sizes) to be detected.**

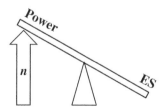

 5. **Statistical software or websites perform Power and Sample Size calculations that can be used to determine the minimum Sample Size required for a given level of Power or Effect Size.**

Explanation

 > 1. **Measurement** (aka Continuous) **data contain more information than Count data, so smaller Sample Sizes can achieve good accuracy.**

The Part 1 article addressed minimum Sample Size for Count data, also known as Discrete data. Count data are non-negative integers: 0, 1, 2, 3, etc. Proportion is the Statistic commonly used with Count data.

Measurement/Continuous data are collected by measuring (widths, for example), not counting. Measurement data will often have decimal points. For example, the following data were collected on the diameters of holes drilled in a manufacturing operation in centimeters: 1.9, 2.1, 2.0, 2.1, 2.1, 2.0, 2.2, 2.0, 2.0, 2.2 cm. The quality specification limits state that the hole must be 2.0 cm $\pm$ 0.1 cm.

If you capture Sample data as measurements, don't convert it to Count data. **Information is lost in converting Measurement data to Count data.**

In the Sample of 10 measurements above, 2 of the measurements were outside the specification limits. So, we could record a Count of 2 defective items in a Sample of size $n = 10$. What we would then know about the Sample would be summarized in one Statistic:

- Proportion (of defects) = 0.2

But if we used the measurements we collected, we'd have a lot more, and more detailed, information at our disposal:

- Mean = 2.06
- Mode = 2.00
- Standard Deviation = 0.097

We could also calculate a number of other Statistics, including Skewedness, Kurtosis, etc. And we can verify that the Sample data are roughly Normally distributed.

Standard Deviation is particularly valuable, because it tells us how much Variation there is in the data. And, as we'll see later, Variation (in the form of the Population or Process Standard Deviation) is a key factor in determining minimum Sample Sizes for Measurement data.

So, Measurement data provide us more statistical information, enabling us to fine-tune our calculations for minimum Sample Size. This enables us to have **smaller Sample Sizes for Measurement data.**

> **2. If your focus is on avoiding an Alpha Error (False Positive), you can calculate minimum Sample Size for tests of the Mean using:**
> - α, the selected Level of Significance
> - σ, Standard Deviation of the Population or Process (or an estimate of it)
> - **MOE,** the desired Margin of Error

Inferential Statistical Studies – those which use information from a Sample to estimate something about the overall Population or Process – are subject to Alpha Errors (False Positives) and Beta Errors (False Negatives). See the article *Alpha and Beta Errors* for more on this.

An Alpha Error is the error of concluding that there is a Statistically Significant difference, change, or effect, when, in reality, there is not. Many behavioral and social science studies focus only on avoiding Alpha Errors. Whether or not this is always appropriate can be debated, but if that is the focus, the minimum Sample Size calculations can be simpler. They can be done using

- α, the selected Level of Significance
- σ, Standard Deviation of the Population or Process (or an estimate of it)
- **MOE,** the desired Margin of Error

We can do this by algebraically manipulating a formula which has these Variables. For example:

$$\text{MOE} = \frac{\sigma \, (\text{critical value})}{\sqrt{n}}$$

σ is the Population or Process Standard Deviation. Often, we don't know it, so we must use an alternative, for example, estimate from previous studies, from an industry standard, or using our Sample Standard Deviation.

Alpha is used to determine the Critical Value. For example, from a table or from software we know that the Critical Value of the Test Statistic z for a 2-sided test with $\alpha = 5\%$ is $z_{\alpha/2} = 1.96$.

A little algebraic manipulation gives us:

$$n = \frac{\sigma^2 \, (\text{Critical value})^2}{\text{MOE}^2}$$

Minimizing False Negative Errors as well as False Positive Errors makes for more complicated Power and Sample Size calculations, as discussed in Key to Understanding #5.

3. All other things being equal, an increase in Sample Size (n) reduces all types of Sampling Errors, including Alpha and Beta Errors and the Margin of Error.

A Sampling "Error" is not a mistake. It is simply the reduction in accuracy to be expected when one makes an <u>estimate</u> based on a portion – a Sample – of the data in Population or Process. There are several types of Sampling Error.

Two types of Sampling Errors are described in terms of their Probabilities:

- *p* **is the Probability of an Alpha Error**, the Probability of a False Positive.
- *β* **is the Probability of a Beta Error**, the Probability of a False Negative.

A third type, the Margin of Error (MOE) is the width of an interval in the units of the data. It is half the width of a 2-sided Confidence Interval.

All three types of Sampling Error are reduced when the Sample Size is increased.

This makes intuitive sense because a very small Sample is more likely to not be a good representative of the properties of the larger Population or Process. But, the values of Statistics calculated from a much larger Sample are likely to be very close to the values of the corresponding Population or Process Parameters.

 4. A larger Sample Size enables smaller differences (Effect Sizes) to be detected.

In manufacturing quality studies, for example, the specification limits which define the acceptable range of measured values can be quite narrow. So, it is important to be able to detect small differences, that is, the Effect Size (ES) of the test is small.

To detect small Effect Sizes, the Power of the test (see the article *Power*) must be high. One way to increase Power is to increase $α$, the acceptable threshold for the Probability of an Alpha Error. But we normally don't want to do that. The only other way to increase Power is to increase Sample Size, n.

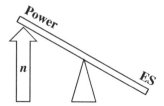

 | **5. Statistical software or websites perform Power and Sample Size calculations that can be used to determine the minimum Sample Size required for a given level of Power or Effect Size.**

We said earlier that, if all you're interested in is avoiding Alpha Errors, then you can use some simple formulas for minimum Sample Size. But for some areas like medical research and process quality improvement, avoiding Beta Errors is very important.

When we add the need to manage Beta Errors and Effect Sizes, the formulas for minimum Sample Size can get very complicated and are beyond the scope of this book. However, there are a number of commercial and free statistical software packages and websites that will do the calculations for you.

Related Articles in This Book: *Sample Size – Part 1: Proportions for Count Data*; *Alpha and Beta Errors*; *Margin of Error*; *Power*

SAMPLING DISTRIBUTION

Summary of Keys to Understanding

 1. The concept of Sampling Distribution is involved in the concepts of the Central Limit Theorem and Standard Error.

 2. The values which comprise a Sampling Distribution are not data values, but Statistics (e.g., the Means) **of Samples.** The Samples must be all of the same Size (*n*), and be taken, With Replacement, from one Population or Process.

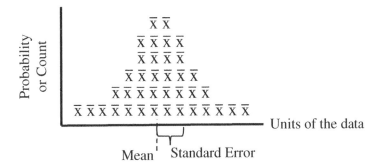

Units of the data

Mean   Standard Error

	Probability Distribution (e.g., Normal, Binomial)	Sampling Distribution
Comprised of:	data values	Statistics
Term for the average	Mean	Expected Value of the Mean
Term for the Standard Deviation	Standard Deviation	Standard Error
Mean and Standard Deviation are expressed in units of:	the data values	the data values

 3. If the Statistic comprising the Sampling Distribution is the Mean or a Proportion, the Central Limit Theorem applies.

Explanation

> **1. The concept of Sampling Distribution is involved in the concepts of the Central Limit Theorem and Standard Error.**

So, some authors say that you need a good understanding of the concept of the Sampling Distribution in order to understand these other two concepts. That can be debated. Sampling Distribution is an abstract concept, and the other two are more concrete and can be understood directly. (See the articles *Normal Distribution* and *Standard Error*.)

In any event, here is how they relate to Sampling Distribution:

- The Standard Error is defined as the Standard Deviation of the Sampling Distribution.

- Central Limit Theorem: Even if the data are not Normally distributed, the Sampling Distribution of the Means or Proportions of the data approaches the Normal Distribution as the Sample Size, n, increases.

> **2. The values which comprise a Sampling Distribution are not data values, but Statistics (e.g., the Means) of Samples. The Samples must be all of the same Size (n), and be taken, With Replacement, from one Population or Process.**

The Distributions with which we are most familiar are Probability Distributions or Frequency Distributions. The horizontal axis is an x Variable such as height, test score, or defect count. The vertical axis is the Probability or Count associated with the value of each x.

In a Sampling Distribution, the horizontal axis represents the value of a Statistic calculated from Samples.

How to create a Sampling Distribution:

Take a number of Samples – With Replacement – of equal size, n – from a Population or Process. ("With Replacement" means that, after collecting a Sample of data, those data values remain in – or are returned to – the Population or Process before we take the next Sample.)

For each Sample, calculate a Statistic – the same Statistic for each Sample, e.g., the Mean of each Sample.

The collection of the values of these Sample Statistics – e.g., all the Sample Means – form a Distribution, called the Sampling Distribution.

In the illustration below, each $\bar{x}$ represents the Mean of one Sample. They are placed above the x axis above the point which represents their value. If more than one occurs within the same narrow range of values, they are stacked one above the other. The height of each stack is proportional to the Probability of that value of the Mean.

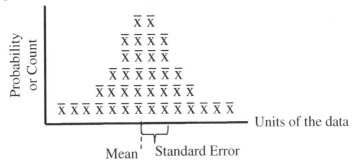

The collection of Means is itself a Distribution, a Sampling Distribution. This is not a Distribution of individual data values, but a Distribution of Statistics (Means) each calculated from a Sample of data.

We can calculate the usual descriptive Statistics for this Sampling Distribution. It has a Mean (called the Expected Value of the Mean) and a Standard Deviation (called the Standard Error of the Mean).

A Sampling Distribution vs. the Sampling Distribution

- We can have a Sampling Distribution produced from a few Samples, as shown in the chart above.
- The Sampling Distribution includes all possible Samples. So, it is a something of a theoretical concept.

The definition of Standard Error refers to the Sampling Distribution.

The Central Limit Theorem is intended to be used with practical numbers of Samples, so it is usually stated with a reference to a Sampling Distribution.

The following table compares and contrasts a Sampling Distribution with the Probability Distributions (e.g., Binomial, Normal, F) with which we are more familiar.

	Probability Distribution (e.g., Normal, Binomial)	**Sampling Distribution**
Comprised of:	data values	Statistics
Term for the average	Mean	Expected Value of the Mean
Term for the Standard Deviation	Standard Deviation	Standard Error
Mean and Standard Deviation are expressed in units of:	the data values	the data values

| **4. If the Statistic comprising the Sampling Distribution is the Mean or a Proportion, the Central Limit Theorem applies.**

There are various wordings of the theorem, but here is the essence:

Central Limit Theorem (CLT):
No matter the shape of the Distribution of the underlying data, if you take multiple Samples (of the same Size, *n*) and compute the Means (or Proportions) for each Sample,

- **the resulting Sampling Distribution of Sample Means (or Proportions) will be approximately Normal,**
- and the Mean (or Proportion) of the Distribution will approximate the Mean (or Proportion) of the Population or Process.
- The larger the Sample Size, the closer these approximations will be. But even relatively small Samples will demonstrate these characteristics.

Related Articles in This Book: *Distributions – Part 1: What They Are*; *Standard Error*; *Standard Deviation*; *Normal Distribution*

SIGMA

Keys to Understanding

 1. Sigma (σ) is a Greek letter which represents the Standard Deviation of a Population or Process. "A sigma" is one Standard Deviation.

(The Roman letter s is used for the Standard Deviation of a Sample.)
See the article *Standard Deviation*.

 2. According to the Empirical Rule for Normal Distributions, roughly **68% of the data lie within 1 Sigma on either side of the Mean, 95% within 2 Sigma, and 99.7% within 3 Sigma.**

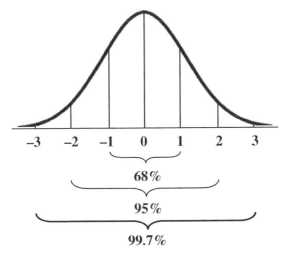

Standard Deviations

See the article *Normal Distribution*.

 3. "Six Sigma" is the name of **a process improvement discipline which aims to have 6 Sigmas define the percentage of defects. Six Sigma translates to 3.4 defects per million opportunities** (for a defect).

See the article *Process Capability Analysis* for specifics of how this translation is done.

(The "7 Sigma" referenced in the dedication to this book would be fewer than 2 defects per 100 million opportunities.)

SKEW, SKEWNESS

Summary of Keys to Understanding

 1. **Skew is a property of the Shape of a Distribution or a Sample of data. Skewness is a measurement of Skew.**

Properties of a Distribution

Central Tendency	Variation/Spread	Shape
Mean	Range	**Skew**
Median	Variance	Kurtosis
Mode	Standard Deviation	

 2. In Statistics, the Skew direction is the opposite of that in every-day language. **The Skew is in the direction toward which the long tail of the Distribution points.**

Skewed Left,
Negative Skew

Skewness = 0

Skewed Right,
Positive Skew

 3. **There are rules of thumb for what the Skew can tell you about the relationship between Mean, Mode, and Median.** But these don't apply for all Distributions.

 4. There are many different types of Skewness with many different formulas and names and symbols. **Understand what type of Skewness measure your software has used for your specific data set, and describe that measure when reporting your results.**

Explanation

 | **1. Skew is a property of the Shape of a Distribution or a Sample of data. Skewness is a measurement of Skew.** |

There are three Categories of measurements which describe Distributions of data:

Properties of a Distribution

Central Tendency	Variation/Spread	Shape
Mean	Range	**Skew**
Median	Variance	Kurtosis
Mode	Standard Deviation	

- Central Tendency: e.g., Mean, Median, Mode
- Variation (aka Variability, Dispersions or Spread): e.g., Range, Standard Deviation
- Shape: e.g., Skewness, Kurtosis

We can say that a Distribution is "skewed to the left" or "skewed to the right." Usually, the term "Skewness" is used to refer to the measurement of Skew. As such, it would be a Parameter for a Population or Process and a Statistic for a Sample.

 | 2. In Statistics, the Skew direction is the opposite of that in everyday language. **It's the direction in which the long tail of the Distribution points.** |

In this article, we'll use the term "Distribution" to include any collection of data. It could be a Population, a Process, or a Sample.

When looking at a curve like the Distribution below, most people would focus on the bulk of the area under the curve, which is to the left. They would say it was "skewed to the left."

Everyday language:
"Skewed to the left"

Statistics:
"Skewed to the right",
"Positive Skew."

But in statistics, we say that it is "Skewed to the right." And since positive numbers on a graph are to the right, we say it "Positively Skewed." The value of its Skewness measurement would be positive.

<u>Memory Cue</u>: Think of "**the tail wagging the dog**". It's the tail that defines the direction of the Skew, not the bulk of the dog's body.

It may also help to think of the Distribution as being "stretched out" in the direction of the Skew.

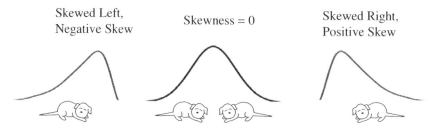

Skewed Left,
Negative Skew

Skewness = 0

Skewed Right,
Positive Skew

The Skew is in the direction of the long tail.

> **3. There are rules of thumb for what the Skew can tell you about the relationship between Mean, Mode, and Median.** But these don't apply for all Distributions.

For example, some books say that,

> for <u>Left-Skewed</u> Distributions: Mean < Mode < Median
> for <u>Right Skewed</u> Distributions: Median > Mode > Mean

Looking at simple shapes like those pictured above, this might make intuitive sense – at least the relationship between the Mean and the Mode.

But Distributions can come in all sorts of weird shapes for which these rules of thumb are not true. Even a simple bimodal (double-humped) Distribution can be a counterexample.

> **4.** There are many different types of Skewness with many different formulas and names and symbols. **Understand what type of Skewness measure your software has used for your specific data set, and describe that measure when reporting your results.**

In other articles, we have noted that there are several different names and symbols for one thing. And we tried to list them all. For Skewness, this appears to be an impossible task.

The noted statistician, Karl Pearson, alone had at least three different formulas for different measures of Skewness. And, unfortunately, a book may refer to "Pearson's Coefficient of Skewness" without specifying which one. There are also a "Percentile Coefficient of Skewness," a "Quartile

Coefficients of Skewness," an "L-skewness," and various "Moment Coefficients of Skewness."

Symbols for Skewness include Sk_p, Sk_q, k, μ_3, b_1, G_1, skew(x), dSkew(x), and more.

Also, different formulas for Skewness are used with different types of Distributions

Since the formulas for Skewness can involve calculations that are cumbersome to do manually (involving summations of differences, for example), software is usually used. The caveat here is: **Understand what type of Skewness measure your software has used for your specific data set, and describe that measure when reporting your results.**

Related Article in This Book: *Distributions – Part 1: What They Are*

STANDARD DEVIATION

Summary of Keys to Understanding

 1. **Standard Deviation is probably the most common measure of Variation in statistics. It is in units of the data.**

 2. **It is the square root of the Variance.**

Population or Process: $\sigma = \sqrt{\dfrac{\sum (x_i - \mu)^2}{N}}$ Sample: $s = \sqrt{\dfrac{\sum (x_i - \bar{x})^2}{n - 1}}$

 3. **Data and Distributions are often described by their Mean and their Standard Deviation.**

 4. **"A Standard Deviation" can itself be used as a unit of measure of Variation.**
   For example, "This data point is 1.5 Standard Deviations from the Mean."

 5. **For both Normal and non-Normal Distributions, each unit of Standard Deviation corresponds to a given Cumulative Probability of the data points**

Standard Deviations	Cumulative % of Data Points within that number of Standard Deviations from the Mean	
	Normal	non-Normal
1	68.27%	NA
2	95.45%	at least 75%
3	99.73%	at least 88.9%
4		at least 93.7%

Explanation

 | 1. **Standard Deviation is probably the most common measure of Variation in statistics. It is in units of the data.** |

There are a number of measures of Variation (also known as "Variability," "Dispersion," and "Spread.")

- Range gives you the upper and lower bounds. But it doesn't tell you anything about how much (if at all) data are clustered around the Mean.
- Variance does tell you about the clustering, but its units are squares of the data. ("The Variance is 2 square kilograms.")
- Mean Absolute Deviation gives you the average distance of data points from the Mean. But its formula uses Absolute values, which are not as conducive to mathematical manipulation as are the squares and square roots used in Variation and Standard Deviation.

Standard Deviation does tell you about clustering. It is also an approximation of the average distance from the Mean. Extreme Outliers can have a disproportionate effect on Standard Deviation.

The units of Standard Deviation are the same as the units of the data. For example if we are collecting data on household incomes in dollars for a community, we may find that the Standard Deviation of incomes is 40,000 dollars.

Notation:
Standard Deviation of a Population or Process: σ ("Sigma")
Standard Deviation of a Sample: s
Variance of a Population or Process: σ^2
Variance of a Sample: s^2

 | 2. **It is the square root of the Variance.** |

Population or Process: $\sigma = \sqrt{\dfrac{\sum (x_i - \mu)^2}{N}}$ Sample: $s = \sqrt{\dfrac{\sum (x_i - \bar{x})^2}{n-1}}$

In the numerators of both these formulas, we have the sum of the squared differences (Deviations) from the Mean. But the denominators are different. The Population Standard Deviation divides by its Size, but the Sample Standard Deviation divides by the Sample Size minus 1. This gives us a somewhat larger value for Sample Standard Deviation. How much larger is dependent on the Sample Size:

n	$n-1$	% difference
5	4	20%
100	99	1%

The $n-1$ adjustment accounts for the fact that small Samples will give less accurate estimations of the Population or Process Standard Deviation than will large Samples. So, **Standard Deviations calculated from small Samples will be wider than those calculated from large Samples.**

 | **3. Data and Distributions are often described by their Mean and their Standard Deviation.**

There are three categories of Statistics/Parameters which are used to describe data and Distributions:

- Central Tendency: e.g., Mean, Mode, Median
- Variation: e.g., Standard Deviation, Variance, Range
- Shape: e.g., Skew, Kurtosis

Much of the data in the world are approximately Normally distributed, and we don't hear much about Skew and Kurtosis. So, describing data or a Distribution by the Mean and Standard Deviation usually gives us a good description of the situation.

Too often, Standard Deviation is not provided in the popular media, so we are left wondering whether the Distribution looks like the one on the left or the right.

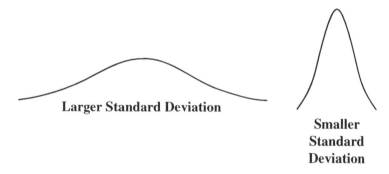

Larger Standard Deviation

Smaller Standard Deviation

 | **4. "A Standard Deviation" can be used as a unit of measure of Variation.**

For example, "This data point is 1.5 Standard Deviations from the Mean."

> **5. For both Normal and non-Normal Distributions, each unit of Standard Deviation corresponds to a given Cumulative Probability of the data points.**

	Percent of Values Found within this Number of Standard Deviations from the Mean			
	1	**2**	**3**	**4**
Normal Distribution (Empirical Rule)	68.27%	95.45%	99.73%	
All Distributions (Chebyshev's Theorem)		>75%	>88.9%	>93.7%

The Normal percentages shown illustrate the **Empirical Rule**, sometimes called the "68, 95, 99.7" Rule.

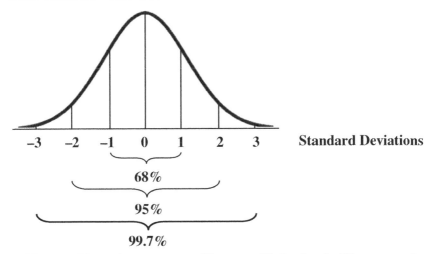

The non-Normal percentages illustrate **Chebyshev's Theorem**, also known as the Bienayme–Chebyshev Rule. The Theorem calculates the percentages with the formula:

$$100\% \times (1 - (1/k)^2)$$

where k is the number of Standard Deviations.

Related Articles in This Book: *Variation/Variability/Dispersion/Spread*; *Variance*; *Normal Distribution*

STANDARD ERROR

An understanding of the concept of Standard Deviation is assumed.

Summary of Keys to Understanding

 1. **Standard Error is the Standard Deviation of a Sample Statistic.**

 2. **Standard Error can be found in formulas in Inferential Statistics. And there are different formulas for different types of Standard Errors.**

Margin of Error = Standard Error × Critical Value

$$\text{SEM} = \frac{s}{\sqrt{n}} \quad \text{SE}(\bar{x}_1 - \bar{x}_2) = \sqrt{\frac{s_1^2}{n_1} + \frac{s_2^2}{n_2}} \quad \text{SE}_p = \sqrt{\frac{p(1-p)}{n}}$$

 3. **Standard Error and Standard Deviation have similarities and differences.**

 4. **Standard Error is shown in the output from Regression analysis and other tests.**

	Coefficients	Std Error	t-Stat	p-Value	Lower 95%	Upper 95%
Intercept	−38.824	32.929	−1.179	0.304	−130.248	52.601
Bedrooms	83.725	11.602	7.217	0.002	51.514	115.937
Bathrooms	76.078	15.469	4.918	0.008	33.129	110.027

 5. **Definition:** The Standard Error is **the Standard Deviation of the Sampling Distribution of a Statistic.**

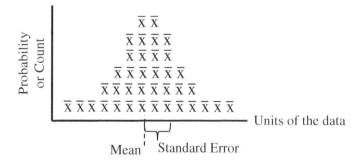

Explanation

> **1. Standard Error is the Standard Deviation of a Sample Statistic.**

First of all, **the "Error" in Standard Error does not imply that a mistake has been made.** It's just a way of saying that any estimate from a Sample can be expected to be less than 100% accurate.

In Inferential Statistics, we collect a Sample of data from a Population. (Note: this book usually says "Population or Process." But that would be repeated many times in this article, so we'll just say "Population" in the rest of this article.) Then, we calculate a Statistic from the Sample data, e.g., the Sample Mean or Standard Deviation. We use the Sample Statistic as an estimate of the corresponding property in the Population, e.g., the Population Mean or Standard Deviation.

In calculating the Standard Error, the Sample data are not used directly. **In place of the x's in one Sample, the Standard Error of the Mean theoretically uses the Means of all possible Samples of the Sample Size** n. The Standard Error is the Standard Deviation of all these Means. (These are the Means in the so-called Sampling Distribution, which is a theoretical concept described in the article by that name.)

Of course, we could never collect data on all possible Samples, but statisticians have derived formulas for calculating the Standard Errors of various Statistics. Here's the one for Standard Error of the Mean, SEM:

$$\text{SEM} = \frac{s}{\sqrt{n}}$$

> **2. Standard Error can be found in formulas in Inferential Statistics. And there are different formulas for different types of Standard Errors.**

To a considerable extent, **Standard Error** plays a behind-the-scenes role. It **is used more frequently as an interim step in calculations or as a component of formulas** than it is as a quoted Statistic in its own right. Experts disagree on whether it is important to understand Standard Error as a separate concept. It is an abstract concept, and it can be confusing.

The *Margin of Error* article in this book shows the following formula:

$$\text{MOE} = \frac{s(\textbf{Critical Value})}{\sqrt{n}}$$

Using the formula for SEM above, we could rewrite this as

Margin of Error = Standard Error × Critical Value

In the article *t-tests – Part 2*, we show the following formula for *t* for a 1-Sample *t*-test. We see the SEM in the denominator.

$$t = \frac{\bar{x} - \mu}{s / \sqrt{n}}$$

That article also show the following generic formula for the Test Statistic, *t*. It applies to all three *t*-tests (which have different kinds of "differences between two Means" and different kinds of Samples). In this case, the concept of Standard Error does simplify the description of what's happening in the three tests.

$$t = \frac{\text{difference between two Means}}{\text{Standard Error}}$$

That article also shows three different formulas for Standard Error, corresponding to the three different *t*-tests. These are in the denominators below:

1-Sample *t*-test: $\dfrac{\bar{x} - \mu}{s / \sqrt{n}}$ 2-Sample *t*-test: $\dfrac{\bar{x}_1 - \bar{x}_2}{s_p / \sqrt{\dfrac{1}{n_1} + \dfrac{1}{n_2}}}$

Paired *t*-test: $\dfrac{\bar{d} - 0}{s_d / \sqrt{n}}$

Another commonly used formula is the Standard Error of the Proportion:

$$\text{SE}_p = \sqrt{\frac{p(1 - p)}{n}}$$

 3. Standard Error and Standard Deviation have similarities and differences.

	Standard Deviation	Standard Error
It is	a Sample Statistic or Population Parameter	a Sample Statistic
Can calculate it for	a Population or a Sample	Sample or Samples
It measures:	Variation	Variation
Variation of what?	individual data values	Statistics (e.g., Sample Means) calculated from Samples
It is the Standard Deviation of	a Sample or a Population	the Sampling Distribution of the Statistic
How calculated	the square root of the Variance	(varies by the Statistic used)

4. Standard Error is provided in the output from Regression analysis and other tests.

	Coefficients	Std Error	*t*-Stat	*p*-Value	Lower 95%	Upper 95%
Intercept	−38.824	32.929	−1.179	0.304	−130.248	52.601
Bedrooms	83.725	11.602	7.217	0.002	51.514	115.937
Bathrooms	76.078	15.469	4.918	0.008	33.129	110.027

Here, the Standard Error is the Standard Deviation of the three Variables in the first column. That may be of interest, but the deciding information is the *p*-value and the 95% Confidence Interval.

 5. Definition: The Standard Error is **the Standard Deviation of the Sampling Distribution of a Statistic.**

Normally one might start with the definition. But this one is so abstract that it probably inhibits, rather than helps, understanding. If you've understood everything up to this point, it may not be worth worrying about trying to make sense of this definition. If you do wish to proceed, please first read the article *Sampling Distribution.*

If we took a lot of Samples – all of the same size, *n* – we could use the Standard Deviation of those Sample Means (from the Population Mean) as a measure of how good an estimate any one Sample is likely to be. The more Samples we took, the more reliable this Standard Deviation could be expected to be.

Theoretically, if we were able to take all possible Samples of a given Sample Size (*n*) and calculate the Standard Deviation of their Statistics, we would calculate the Standard Error. **The Distribution of a specified Statistic of all possible Samples of equal size *n* is called the Sampling Distribution.** (If a Sampling Distribution does not include all possible Samples, it is called a Sampling Distribution.)

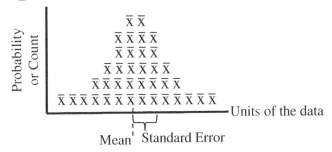

Above is a conceptual illustration of a Sampling Distribution. The usual Probability Distributions (say, the Normal, Exponential, or Binomial) have the x values of the data represented along the horizontal axis. In a Sampling Distribution, however, the Statistic (e.g., the Mean) of each Sample is graphed instead of a data point.

Related Articles in This Book: *Standard Deviation*; *Inferential Statistics*; *Margin of Error*; *Critical Value*; *Confidence Intervals – Part 1*; *Hypothesis Testing – Parts 1 and 2*; *Null Hypothesis*; *t-Tests – Part 1: Overview*; *Variation/Variability/Dispersion/Spread*; *Sampling Distribution*

STATISTICALLY SIGNIFICANT

Summary of Keys to Understanding

 1. A difference, change, or effect which is observed in Sample data is considered "Statistically Significant," if there is a high Probability that the difference, change, or effect is real for the whole Population or Process from which the Sample was taken.

 2. Stated another way, **a difference, change, or effect** which is observed in Sample data **is considered "Statistically Significant," if there is a low Probability that the difference, change, or effect is not real.** That is, there is a low Probability of an Alpha Error (False Positive).

 3. p is the Probability of an Alpha Error. The following things make this Probability lower (and the Probability of Statistical Significance higher).
- **a bigger difference, change, or effect shown in the Sample data**
- **a bigger Sample Size**
- **smaller Variation** (Standard Deviation) **in the Sample**

 4. The person performing the statistical analysis defines what it means to be Statistically Significant by selecting a value for the clip level Alpha (α), the Level of Significance.

 5. There is a Statistically Significant difference, change, or effect if $p \le \alpha$.

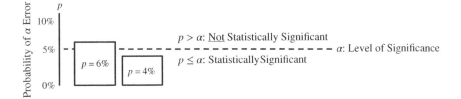

357

Explanation

> **1. A difference, change, or effect which is observed in Sample data is considered "Statistically Significant," if there is a high Probability that the difference, change, or effect is real** for the whole Population or Process from which the Sample was taken.

In Inferential Statistics, we collect a Sample of data from a Population or Process. Then we perform a statistical analysis on the data. We use the results of that analysis to make a conclusion about the Population or Process from which the Sample was drawn. For example,

- We poll 50 people exiting a voting location and determine that Candidate A was preferred 52% to 48%. Is this a Statistically Significant difference? That is, is there a high Probability that Candidate A being in the lead is true for the Population as a whole?

- A drilling Process has historically had a Standard Deviation of 0.010. Our latest sampling indicates a Standard Deviation of 0.012. Is this a Statistically Significant change? That is, is there a high Probability that the Standard Deviation has, in fact, changed?

- A new software program for handling insurance claims was tested with 20 claims and found to reduce the Mean processing time by 5 minutes. Is this a Statistically Significant effect? That is, is there a high Probability that rolling out this new software will improve the operation company-wide?

Q: We keep saying "high Probability," but how high is high enough?
A: Higher than our chosen Level of Confidence. **If we want a 95% Level of Confidence that the difference, change, or effect is real, we would select a 5% Level of Significance** ($\alpha = 5\%$). Our selection of this Level of Significance defines what is Statistically Significant and what is not. (More on this in Key to Understanding #3.)

> **2.** Stated another way, **a difference, change, or effect** which is observed in Sample data **is considered "Statistically Significant"** if there is a **low Probability that the difference change, or effect is not real.** That is, there is a low Probability of an Alpha Error (False Positive).

I saw a unicorn.

Alpha Error
(False Positive)

An Alpha Error (aka Type I Error) is the error of seeing something that is not there.

It **is the error of concluding** – from Sample data – **that there is a difference, change, or effect in the entire Population or Process as a whole, when, in fact there is not.**

A common example: in a medical diagnostic test, a "positive" indicates that a disease is present. If the test conclusions are wrong, and there is no such problem, it would be a False Positive (an Alpha Error).

> **3.** p **is the Probability of an Alpha Error. The following things make this Probability lower (and the Probability of Statistical Significance higher):**
> - **a bigger difference, change, or effect shown in the Sample data**
> - **a bigger Sample Size**
> - **smaller Variation** (Standard Deviation) **in the Sample**

All this, thankfully, makes common sense.

- **a bigger difference, change, or effect in the Sample data**
  Let's say we are comparing the heights of adult males with Samples from two countries. If the difference of the averages of the two Samples is 1 cm, we might wonder whether that's a real difference in the Populations, or if it's just due to random Variation in selecting subjects for the Samples.

  However, if the difference of the averages is 5 cm, then we would intuitively feel more confident that the difference is real.

- **a bigger Sample Size**
  One would expect a Sample with more items to be a more accurate representation of the overall Population or Process. This is true.

- **smaller Variation** (Standard Deviation) **in the Sample**
  Let's say we're comparing the Mean of test scores in our school (84) with the national average of 80. And, let's say that most of the scores

in our school are clustered between 81 and 87, with a high of 95 and a low of 75. This shows a relatively small amount of Variation, and it would tend to argue that the difference is real.

On the other hand, if our scores were spread out from 50 to 100, with very little clustering about the Mean, that would be a less compelling case for a real difference.

p **is the key decision-making output from Inferential Statistical analyses, like the *t*-tests, ANOVA, and Regression.** This is explained in Key to Understanding #5.

> **4. The person performing the statistical analysis defines what it means to be Statistically Significant by selecting a value Alpha (α), the Level of Significance.**

OK, so we know there's going to be some chance, *p*, of an Alpha Error in our testing. How much are we willing to tolerate? How high a value of *p* will we accept before we say the observed difference, change, or effect is not Statistically Significant?

The answer varies by the situation being analyzed. There's more information on this question in the articles *Alpha, α* and *Alpha and Beta Errors*. **The person performing the analysis gets to choose the upper limit for *p* which defines the difference between Statistically Significant and Not Statistically Significant.**

This upper limit or clip level for *p* (which is the Probability of an Alpha Error) **is called Alpha (symbol α).** It is important to select the level for Alpha prior to collecting Sample data. Otherwise, if we take a peek at the data, it might influence our choice for Alpha. And that would corrupt the validity of our analysis.

Most of the time, 5% is selected. So, let's use that going forward. Conceptually, it may help to think in terms of the Level of Confidence first and have that determine our Level of Significance (α). **The Level of Confidence = 1 – α** (in percentages, 100% – α). So, if we want to be 95% confident of not having a False Positive error, then we would select Alpha to be 5%.

I want to be 95% confident of avoiding an Alpha Error.

So, I'll select $\alpha = 5\%$.

Why wouldn't we always select Alpha to give us a 99.99% Level of Confidence? This is explained in the article *Alpha and Beta Errors*. But briefly, the lower the level of Alpha Error we select, the higher the level of Beta Error ("False Negative") we must be willing to tolerate.

> **5. There is a Statistically Significant difference, change, or effect if $p \leq \alpha$.**

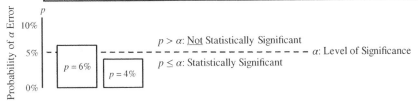

In the diagram above, the clip level Alpha is identified by the dotted line separating Not Statistically Significant from Statistically Significant. One can see why Alpha is called the Level of Significance. Any *p*-value at that level or below indicates Statistical Significance.

Any value for *p* (which is calculated from the Sample data) **less than or equal to Alpha leads us to conclude that the difference, change, or effect observed in the Sample is Statistically Significant.** (Remember *p* is the Probability of an Alpha Error. And Alpha was selected as the boundary value for that Probability separating Statistically Significant from not Statistically Significant.)

If $p \leq \alpha$, then any difference, change, or effect in the Sample <u>is</u> Statistically Significant.

If $p > \alpha$, then any difference, change, or effect in the Sample is <u>not</u> Statistically Significant.

So, the 6% pictured above is in the Not Statistically Significant range. But the 4% value for *p* is in the Statistically Significant range. (5% is also in that range, since that range includes the line).

For many of us, it may be difficult at first to remember which way the "<" points. Is it $p < \alpha$ or $p > \alpha$ which indicates Statistical Significance? One thing which might help is to remember that *p* **represents error, so we would like it to be small before we accept a conclusion of Statistically Significant.** In addition, this book has a separate article – *p, t, and F: "<" or ">"? –* with three different tips for remembering which is which.

In Hypothesis Testing

> $p \leq \alpha$ **causes us to Reject the Null Hypothesis.** The Null Hypothesis states that there is no (Statistically Significant) difference, change, or effect. $p \leq \alpha$ says that there is.

$p > \alpha$ **causes us to Fail to Reject** (i.e., to Accept) **the Null Hypothesis** that there is no Statistically Significant difference, change, or effect.

To summarize some of key points in this article:

	Statistically Significant	**Not Statistically Significant**
There is a high Probability that the difference, change, or effect observed in the Sample is real for the Population or Process as a whole.	Yes	No
Probability of an Alpha Error	$p \le \alpha$	$p > \alpha$
Null Hypothesis	Reject	Accept/Fail to Reject

Related Articles in This Book: *Alpha and Beta Error*; *p, p-Value*; *Alpha, α*; *p, t, and F: "<" or ">"?*

SUMS OF SQUARES

Keys to Understanding

This article is provided as a 1-page summary of the concept. For more information, see the articles, ANOVA – Parts 2 and 3, and Regression – Part 1.

 1. A Deviation is the difference from a single data value to a target value, e.g., the Mean $(x - \bar{x})$ or a point on a Regression line $(y - y_R)$.
Each individual Deviation is squared to eliminate negative numbers, e.g., $(x - \bar{x})^2$, **then all are totaled to give the** Sum of Squared Deviations, or **Sum of Squares (SS).**

$$SS = \sum (x - \bar{x})^2$$

 2. Sums of Squares come in several types. Sums of Squares are measures of Variation. A Sum of Squares divided by its Degrees of Freedom is a Variance (symbol s^2).

$$s^2 = \frac{\sum (x - \bar{x})^2}{n - 1} = \frac{SS}{n - 1}$$

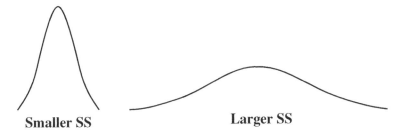

Smaller SS   **Larger SS**

 3. ANOVA uses Sum of Squares <u>Between</u> groups (SSB) and Sum of Squares <u>Within</u> groups (SSW). Their respective Means – **MSB and MSW – are Variances.**

Sum of Squares <u>Total</u>, **SST = SSB + SSW**

$$F = MSB/MSW$$

 4. In Regression, Sum of Squares Total (SST) measures the total Variation in the y Variable. **Sum of Squares Regression (SSR)** is that portion of SST which is explained by the Regression Line. **Sum of Squares Error (SSE)** is what's left over.

$$SST = SSR + SSE$$

t – THE TEST STATISTIC AND ITS DISTRIBUTIONS

Summary of Keys to Understanding

 1. *t* is a **Test Statistic** used in tests involving the **difference between two Means**.

 2.
$$t = \frac{\text{difference between two Means}}{\text{Standard Error}}$$
$$= \frac{(\text{difference between two Means})\sqrt{n}}{s}$$

 3. *t* is a measure of **how likely** it is that a difference in Means is Statistically Significant.

Larger Difference between Means $\Longrightarrow$ Larger *t* $\Longrightarrow$ **More** likely

Larger Sample Size $\Longrightarrow$ Larger *t* $\Longrightarrow$ **More** likely

Larger Sample Standard Deviation $\Longrightarrow$ Smaller *t* $\Longrightarrow$ **Less** likely

⎫ that the difference is Statistically Significant ⎬

 4. **There is** not one *t*-Distribution, but **a different *t*-Distribution for each value of the Degrees of Freedom, df. As df grows larger, the *t*-Distribution approaches the *z*-(Standard Normal) Distribution.**

 5. *t* **has a number of similarities to** *z* **and some key differences.**

 6. **Use *t* instead of *z* when**
 – the Standard Deviation (**Sigma, σ**) of the Population or Process **is unknown**
 – **or the Sample size is small** ($n < 30$)

Explanation

> **1. *t* is a Test Statistic used in tests involving the difference between two Means.**

t is also known as "Student's *t*" or the "*t*-statistic."

... Test Statistic ...

A Statistic is a property of a Sample. A Test Statistic is one that has an associated Probability Distribution (or associated family of Probability Distributions). So, **for any value of the Test Statistic, we can determine the Probability of that value**. More importantly, we know **the Cumulative Probability of all values greater or less than that particular value.** This is an essential part of Inferential Statistics, in which we estimate (infer) a Parameter (e.g., the Mean or Standard Deviation) of a Population or Process based on the corresponding Statistic of a Sample. Common Test Statistics are *t*, *z*, *F*, and χ^2 (Chi-Square).

... the difference between two Means.

One Mean is always the Mean of a Sample.

The Second Mean can be either

- **A specified Mean**, such as a target Mean, a historical Mean or an estimate, or
- **The Mean of a Sample from a different Population or Process** than the first Mean, or
- **A second Mean from the same test subjects** (e.g., before and after some event).

These three different types of the second Mean correspond to three different *t*-tests. See the articles *t-tests – Part 1* and *Part 2*.

t-test	Mean 1	Mean 2
1-Sample	Sample from a Population or Process	Specified Mean (a target, an estimate, or a historical value)
2-Sample	Sample from a Population or Process	Sample from a different Population or Process
Paired	one half of a Sample of paired data, e.g., score before training	the other half of the Sample of paired data for the same test subjects, e.g., score after training

> **2. Descriptive formula:**
>
> $$t = \frac{\text{difference between two Means}}{\text{Standard Error}}$$

The numerator is straightforward enough, but what is the Standard Error? Standard Error is a measure of Variation in the Sample(s). See the article *Standard Error* for more information.

Different *t*-tests have different formulas for the Standard Error. But, a generic formula for Standard Error can be stated as

$$\textbf{Standard Error} = s/\sqrt{n}$$

where *s* is the Standard Deviation of the Sample and *n* is the Sample Size.

Since Standard Error is in the denominator of the formula for *t*, the square root of *n* is in the denominator of the denominator. Let's simplify the formula with simple algebra:

$$t = \frac{(\textbf{difference between two Means})\sqrt{n}}{s}$$

> **3. *t* is a measure of how likely it is that a difference in Means is Statistically Significant.**

As with all Test Statistics, we compare *t* to its Critical Value. The value of *t* is calculated from Sample data, as shown in the formulas above. The value of *t*-critical is determined by the value selected for Alpha, the Significance Level, and the appropriate *t*-Distribution.

A large value for *t* makes it more likely to be larger than *t*-critical, and so makes it more likely that there is a Statistically Significant difference in the Means.

Since the difference between the Means and the Sample Size, *n*, are in the numerator, larger values for either of these would make *t* larger. Since it's actually the square root of *n* that is in the numerator, an increase in the difference between the Means would have much more of an effect than a proportional increase in the Sample Size.

Since the Standard Deviation is in the denominator, a larger Variation in the Sample(s) will make *t* smaller.

This is all summarized in the graphic below:

Larger Difference between Means ⟹ **Larger *t*** ⟹ **More** likely ⎤
                                                  that the

Larger Sample Size ⟹ **Larger *t*** ⟹ **More** likely ⎬ difference is Statistically Significant

Larger Sample Standard Deviation ⟹ **Smaller *t*** ⟹ **Less** likely ⎦

Some authors say that *t* is a measure of "how good the Sample is" or "how accurate the Sample is in estimating the Population or Process Mean." But it's probably more accurate to say **that *t* is a measure of how likely it is that a difference in Means is Statistically Significant.**

> 4. **There is** not one *t*-Distribution, but **a different *t*-Distribution for each value of the Degrees of Freedom, df. As df grows larger, the *t*-Distribution approaches the *z*-(Standard Normal) Distribution.**

For a single Sample, df $= n - 1$, where *n* is the Sample Size. For other situations it gets more complicated, for example, in the 2-Sample *t*-test, df $= n_1 + n_2 - 2$.

A *t*-Distribution is like a *z*-(Standard Normal) Distribution which has been modified to account for the increase in Variation due to being Sample-based.

It stands to reason that the smaller the Sample, the less likely it is to accurately depict the Population or Process. So, **the Standard Deviation of a *t*-Distribution for a small Sample is larger than that for a large Sample**. The Distribution curve for a small Sample would be spread wider than for a larger Sample.

For a (theoretical) Sample Size of infinity, the *t*-Distribution and the *z*-Distribution are identical – the Standard Normal Distribution.

> 5. *t* **has a number of similarities to *z* and some key differences.**

	z	*t*
It is a ...	Test Statistic	
Probability Distribution(s)	Bell-shaped, Symmetrical, Never touches the horizontal axis	
	1	1 for each value of df
Mean = Mode = Median	Yes, and they all = 0	
Formula	$z = (\bar{x} - \mu)/\sigma$	$t = (\bar{x} - \mu)/(s/\sqrt{n})$
Varies with Sample Size	No	Yes
Accuracy	less	more
Standard Deviation	$= 1$	Always > 1

where

> df is Degrees of Freedom; $\bar{x}$ is the Sample Mean; μ is the Population or Process Mean; s is the Sample Standard Deviation; σ is the Population or Process Standard Deviation; n is the Sample Size.

z **and** t **are both Test Statistics.** There is only one z-Distribution, but **there is a whole family of** t**-Distributions,** one for each value of df. For a single Sample (as in a 1-Sample t-test) df $= n - 1$.

The Probability Distributions of both z **and** t **are centered on zero. They are both symmetrically bell-shaped.** In addition, for both z and t, the Probability Distribution curves never touch the horizontal axis. (**The Probability is never zero for any value of** z **or** t.) The left and right tails of the curves just extend to infinity, getting ever closer to zero.

The z-Distribution is the idealized Standard Normal Distribution, with Mean = Mode = Median = 0 and a Standard Deviation of 1.

The key practical difference between z **and** t **is that** t **takes into account the error introduced when Sample Statistics** (e.g., the Sample Mean) **are used as estimates for the Population or Process Parameter** (e.g., the Population or Process Mean). The formula for t includes a term for the Sample Size, n; z's formula does not. So the value of z is the same for a Sample Size of 1 or a Sample Size of 1000. You can see how z would generally be less accurate than t – much less accurate for small Sample Sizes ($n < 30$).

	6. Use t **instead of** z **when**
	– the Standard Deviation (**Sigma,** σ) of the Population or Process **is unknown**
	– or the **Sample size is "small"** ($n < 30$)

The formula for z is $\qquad z = (\bar{x} - \mu)/\sigma$

σ (Sigma) is the Population or Process Standard Deviation. **If you don't know the Population or Process Sigma,** and you rarely do, **you can't use the formula for** z. Instead, you'd have to use an estimate for Sigma, and that would be s, the Standard Deviation of a Sample. But z doesn't take into account the fact that you're using an estimate from a Sample, so its calculation is less legitimate. That's why you need to use t.

As the Sample Size increases, the Probabilities of t **get closer and closer to the Probabilities for** z. For $n =$ infinity, they are the same. So, it is often said that you **can use** z **instead of** t **when the Sample is "large."**

Many experts say 30 is large enough, others say 100. **Some say it's best to use *t* whenever a Sample is used.**

Related Articles in This Book: *t-tests – Parts 1 and 2*; *Standard Deviation*; *Standard Error*; *Alpha, p-Value, Critical Value, and Test Statistic – How they Work Together*; *Degrees of Freedom*; *z*

t-TESTS – PART 1 (OF 2): OVERVIEW

Builds on the content of the article "t – the Test Statistic and Its Distributions".

Summary of Keys to Understanding

 1. The three types of *t*-tests are differentiated by the types of Means that they compare:

	Means being compared
1- Sample	Sample Mean to a specified Mean
2-Sample	Means of Samples from two different Populations or Processes
Paired	Mean of the differences in pairs of measurements to a Mean of zero

 2. Assumptions for the data are: Normality, and – for the 2-Sample *t*-test – equal Variance.

 3. The <u>1-Sample</u> *t*-test compares a calculated Sample Mean to a Mean we specify.

 4. The <u>2-Sample</u> (aka Independent Samples) ***t*-test compares the Means of two Samples <u>from different Populations or Processes.</u>**

 5. The Paired – aka Dependent Samples – ***t*-test compares the differences between pairs of measurements taken <u>from the same test subjects</u>** at different times or under different conditions.

Explanation

Builds on the content of the article "t – the Test Statistic and Its Distributions".

1. The three types of *t*-tests are differentiated by the types of Means that are compared:

	Means being compared
1- Sample	Sample Mean to a specified Mean
2-Sample	Means of Samples from two different Populations or Processes
Paired	Mean of the differences in pairs of measurements to a Mean of zero

t is the Test Statistic for the *t*-tests. **We use *t* when we are trying to determine if there is a Statistically Significant difference between two Means.** (For three or more Means, use ANOVA.)

Each of the three types of *t*-tests compares two Means, but the two Means are of different types for the different tests. This results in three different mathematical formulas for calculating *t* and for the Degrees of Freedom, df. These are described in the Part 2 article. The article *t, the Test Statistic and Its Distributions* explains how there are different *t*-Distributions for each different value of df.

2. Assumptions for the data are: Normality, and – for the 2-Sample *t*-test – equal Variance.

Tests in statistics usually have restrictions on the kinds of data that can use the test. These are called "Assumptions."

| **OK** | **NOT OK** | **NOT OK** |

The data are assumed to be roughly Normal

The *t*-Distributions are <u>bell-shaped and symmetrical</u>. As the Sample Size increases, *t*-Distributions get closer and closer to the Normal Distribution. So, if we are going to use the values and probabilities of a *t*-Distribution, the data must be at least roughly "Normal."

How do we know if the data are "Normal enough"?

Many software packages do a Normality test, such as the Anderson–Darling, as part of the *t*-test. If the Anderson–Darling *p*-value is greater than the chosen Significance Level, Alpha, then the data are "Normal enough." To be certain, it's good to look at a Normal Probability plot, to see if the plot points roughly track the straight line relatively closely.

What do we do if the data are not "Normal enough"?

There are so-called "Non-Parametric" counterparts to "Parametric" tests (those which require roughly Normal data).

The Non-Parametric counterparts to the 1-Sample *t*-test include the 1-Sample sign or Wilcoxon signed rank tests. These use Medians, rather than Means.

For the 2-Samples *t*-test, use the Mann–Whitney test, which also uses Medians.

For the Paired *t*-test, use the Wilcoxon signed-rank test for paired Samples.

For the 2-Sample *t*-test, roughly equal Variance is also required.

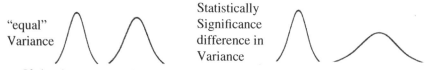

"equal" Variance

Statistically Significance difference in Variance

If the data are roughly Normal, the *F*-test can be used to test for equal Variances.

If the Variances are not roughly equal, Welch's *t*-test can be used.

 | **3. The 1-Sample *t*-test compares a calculated Sample Mean to a Mean we specify.**

In the 1-Sample *t*-test, there is only 1 Sample of data and 1 Sample Mean which is calculated from that Sample. We need to compare two Means; where does the other Mean come from? It can come from anywhere. Some common examples are:

- A Mean calculated outside our test. For example, we compare Mean exam scores for our high school class with the national average.
- An estimated or hypothesized Mean.
- A historical Mean. For example, we may suspect that the Mean of our Process has drifted slightly from what it's always been.

- A target Mean. For example, is the current Mean number of defects in our production Process below our target?

> **4. The 2-Sample** (aka Independent Samples) *t*-**test compares the Means of two Samples from different Populations or Processes.**

In the 2-Sample *t*-test, one Sample of data is taken from each of two different Populations or Processes.

The Means of the two Samples are compared in the *t*-test to determine if there is a Statistically Significant difference.

This is sometimes called the Independent Samples *t*-test. Samples are said to be independent if they come from unrelated Populations or Processes, and the Samples have no effect on each other. For example in a test of a new drug, one Population took the drug and the other Population took the placebo. Different patients – different "experimental units" (aka "statistical units") – took either the drug or the placebo.

The table below illustrates the difference between the 2-Sample and the Paired *t*-test. Note that, **in the 2-Sample** (Independent Samples) *t*-**test, there are different test subjects in Sample 1 and Sample 2. But in the Paired** (Dependent Samples) *t*-**test, the same people are measured twice.**

2-Sample *t*-test				Paired *t*-test		
Sample 1 Not trained $n_1 = 6$		Sample 2 Trained $n_2 = 5$			Before Training	After Training	Difference $n = 5$
J. Black	72	A. Conrad	76	K. Albert	74	78	+4
T. Gerard	80	J. David	78	P. Jacobs	76	83	+7
M. Lowry	78	W. Johns	83	T. Smith	73	81	+8
P. Mason	74	F. Lyons	86	R. Wang	81	84	+3
R. Vargas	79	M. White	61	D. Young	78	86	+8
B. Wilson	70						

> **5. The Paired** aka Dependent Samples *t*-test **compares the differences between pairs of measurements taken from the same test subjects** at different times or under different conditions.

The Paired *t*-test is also called the Dependent Samples *t*-test or the Paired Samples *t*-test. But, these two names can be misleading, because they imply that two Samples are taken.

There are two sets of measurements taken – before and after, in the above example. And we provide the software with the two sets. But **the test treats it as one Sample which consists of the differences** calculated between the two measurements taken from each individual test subject.

The table above right shows 10 measurements of worker production in 5 pairs, and $n = 5$ is the Sample Size. That is the number of pairs, the number of differences. **The Paired *t*-test tells us whether the <u>differences</u> are Statistically Significant.**

But didn't we say that *t*-tests compare two Means? Should we be comparing the Mean before with the Mean after? Well, if we did that, we'd be losing some information – specifically how much Variation there is in the differences for individuals. If some individuals improved greatly and others not at all, there would be a large Standard Deviation in the differences which may not show if we averaged them all together. And, as we'll see in Part 2, the Standard Deviation is involved in calculating the value of *t*.

So, we can calculate a Mean for the differences; what is the other Mean we compare it to? Essentially a specified Mean of zero. So, the Paired *t*-test is like the 1-Sample *t*-test in which the Sample of differences is compared to specified Mean of zero.

The most common use for the Paired *t*-test is for a "before and after" analysis. For example, Does a training program make a Statistically Significant difference in the production output of individual workers? The 2-Sample *t*-test would not work as well for answering that question. There are any number of Factors which can affect worker production output. Experience is one. If one Sample has workers with more experience than the other Sample, then that could give us misleading results. But if we measure the same workers before and after training, we can eliminate other Factors, like experience, so that we can focus on only one Factor – the training.

So, compared to the 2-Samples *t*-test **the Paired *t*-test does a much better job of Blocking out the effect of other Factors.** (See the article *Design of Experiments (DOE) – Part 3* for more on Blocking.) The price to pay for this is that we need to take twice as many measurements to get the same value for Sample Size, *n*. And, all other things being equal, a larger Sample Size gives a more accurate result in our testing.

This article, Part 1, gives an overview of *t*-tests and describes the three different types. Part 2 describes the calculations and analyses that happen when *t*-tests are done.

Related Articles in This Book: *t – the Test Statistic and Its Distributions*; *Degrees of Freedom*; *Normal Distribution*; *Non-parametric*; *F*; *Design of Experiments (DOE) – Part 3*; *t-tests – Part 2: Calculations and Analyses*

t-TESTS – PART 2 (OF 2): CALCULATIONS AND ANALYSIS

Prerequisite: t-tests – Part 1: Overview

Summary of Keys to Understanding

 1. Much the same thing happens in the three types of *t*-tests covered in Part 1. There are just different formulas for *t* and for the Degrees of Freedom, df.

 2. First, identify the Means to be compared and the test type. Next, select a Significance Level, Alpha (α). Then, collect a Sample or Samples of data.

 3. Then, use the Sample data to calculate some Descriptive Statistics, *t*, and the Degrees of Freedom, df.

 4. Use the appropriate *t*-Distribution and Alpha to determine *t*-critical.

 5. Or use the appropriate *t*-Distribution and *t* to determine *p*.

 6. If *t* ≥ *t*-critical (equivalently, if *p* ≤ α), then there is a Statistically Significant difference between the Means. Otherwise, there is not.

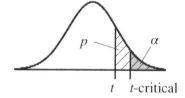

Means <u>are</u> different

Means are <u>not</u> different

Explanation

Prerequisite: t-tests – Part 1: Overview

> **1. Much the same thing happens in the three types of *t*-tests covered in Part 1. There are just different formulas for *t* and for the Degrees of Freedom, df.**

t is the Test Statistic for the *t*-tests. The article in this book, *t, the Test Statistic and Its Distributions* says, **"*t* is a measure of how likely it is that a difference in Means is Statistically Significant."**

Like all Test Statistics, **t is calculated from Sample data.** Here is a descriptive formula:

$$t = \frac{\text{difference between two Means}}{\text{Standard Error}}$$

Each of the three types of *t*-tests compares two Means, but the two Means are of different types for the different tests. So, **the mathematical formula for the numerator is different for each type of *t*-test. Likewise, the mathematical formula for the denominator, the Standard Error, is different for each type of *t*-test.**

And finally, each type of test has a different formula for the Degrees of Freedom, df. df is not involved in the formula for *t*. However, there is a different *t*-Distribution for each different value of Degrees of Freedom. And different Distributions give different values for the Probability of a given value of *t*.

You don't have to memorize these formulas. Spreadsheets or statistical software can do all these calculations for you once you select the *t*-test and provide the data and a value for Alpha, the Level of Significance. However, you may find these formulas useful in getting a conceptual understanding of what's going on in the tests.

t-test	*t*	df
1-Sample	$\dfrac{\bar{x} - \mu}{s/\sqrt{n}}$	$n - 1$
2-Sample	$\dfrac{\bar{x}_1 - \bar{x}_2}{s_p/\sqrt{\dfrac{1}{n_1} + \dfrac{1}{n_2}}}$	$n_1 + n_2 - 2$
Paired	$\dfrac{\bar{d} - 0}{s_d/\sqrt{n}}$	$n - 1$ (*n* is the number of pairs)

$\bar{x}, \bar{x}_1$, and $\bar{x}_2$ are Sample Means.

μ is a specified Mean.

$\bar{d}$ is the Mean of the differences in the two values comprising each pair.

s_p is called the "pooled Standard Deviation"; it has its own multi-term formula.

s_d is the Standard Deviation of the differences.

Note the similarities in the formulas for 1-Sample and Paired *t*-tests. They are similar in concept, except that the Paired test uses calculated differences as the data in its Sample. Usually the formula for *t* in the Paired test is shown using $\bar{x}_1 - \bar{x}_2$ instead of $\bar{d} - 0$. But that can be confusing, since – as we explained in Part 1 – there are <u>not</u> two Samples involved. There is a pair of measurements on each test subject (e.g., a worker's efficiency measurements before and after training). The <u>differences</u> in the two measurements for each individual become the data values in the single Sample.

 | **2. First, identify the Means to be compared and the test type. Next, select a Significance Level**, Alpha (α). **Then, collect a Sample or Samples of data.**

This is the usual sequence in Inferential Statistical analyses. The order is important.

How to Do a *t*-test		
Step	**Input** ⇨ **Process** ⇨	**Output**
A.	The situation to be analyzed / Identify the Means to be compared	two Means identified Null Hypothesis (optional)
B.	Judgment of the tester / Select a Significance Level	Alpha, α (often 0.05)
C.	Population or Process / Collect data	one or two Samples of data

Step A: Identify the Means to be compared and the test type.

One will determine the other; see the Part 1 article.

If Hypothesis Testing is to be used, a Null Hypothesis must be stated at this time. The question or problem to be resolved by the test is stated in the negative. Examples:

- There is <u>no</u> (Statistically Significant) <u>difference</u> between the Means of these two Populations.

- There has been <u>no</u> (Statistically Significant) <u>change</u> in the Mean of this measurement in the Process from its historical Mean.
- The training has had <u>no</u> (Statistically Significant) <u>effect</u>.

Step B: Select a Significance Level, Alpha (α).

As noted in the article in this book, "Alpha, α", in order to keep the integrity of any test, we must select a value for Alpha, the Level of Significance <u>before</u> collecting a Sample of data. **The Level of Confidence is $1 - \alpha$. So, if we want to be 95% confident in our conclusions, we select $\alpha = 5\%$ (0.05).** This is our clip level for the Statistical Significance mentioned in the Null Hypothesis examples above.

Step C: Collect the Sample(s) of data.

3. Then, use the Sample data to calculate some Descriptive Statistics, *t*, and the Degrees of Freedom, df.

Step D: Calculate Descriptive Statistics from the Sample data. For the 2-Sample *t*-test, there will be two sets of $\bar{x}$, *s*, *n*, and **df.**

Step E: Calculate *t*.

Step	Input ⇨	Process ⇨	Output
D.	Sample data	Calculations, counting	Sample Mean: $\bar{x}$ Sample Std. Deviation: *s* Sample Size: *n* Degrees of Freedom: **df**
E.	$\bar{x}, s, n$ 1-Sample test: specify an estimate or target for the Population or Process Mean, μ (if applicable)	Calculate *t* using the formula appropriate to the type of *t*-test	*t*
F.	**df**	Table lookup or software	The *t*-**Distribution** to be used

Step F: Use the Degrees of Freedom, df, to identify the appropriate *t*-Distribution.

A Test Statistic is one which has an associated Probability Distribution or family of Probability Distributions. *t* **has a different Distribution for each different value of df**. As df gets larger and larger, the *t*-Distribution more closely approximates the Standard Normal Distribution (for which z is the Test Statistic).

> ### 4. Use the appropriate *t*-Distribution and Alpha to determine *t*-critical.

Step	Inputs	⇨ Process	⇨ Output
G.	**Alpha,** the *t*-**Distribution,** 1-tailed or 2-tailed	Calculate the boundary of an area under the curve	***t*-critical**

Given the *t*-Distribution and the information on whether it's a 1-tailed or 2-tailed analysis, **the *t*-test uses the Cumulative Probability, Alpha, to calculate the numerical value, *t*-critical.** Critical values of t are available in tables, or they can be calculated with spreadsheets or software.

***t*-critical and Alpha convey the same information**. Either one can be derived from the other along with the *t*-Distribution.

A *t*-Distribution is a Probability Distribution pictured as in the curves below. The horizontal axis is for values of t. The height of the curve over a given value of t is the Probability of that value. Of more interest is the Cumulative Probability of a range or ranges of values of t, because Alpha and p are Cumulative Probabilities. **A Cumulative Probability is calculated as the area under the curve of the range of *t*-values.**

Let's say we have selected $\alpha = 5\%$. So, we shade 5% of the area under the curves below.

1-tailed

For a 1-tailed (right-tailed or left-tailed) test, we shade either the rightmost or left-most 5% under the curve. In the *t*-Distributions pictured below, ***t*-critical is calculated as the value of t which defines the boundary of the shaded area under the curve representing Alpha.**

Right-tailed test: Have our school's tests scores shown a Statistically Significant better performance than the national average?

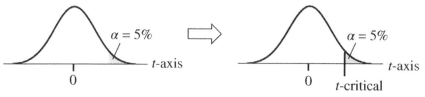

Left-tailed test: Does our new Process show a Statistically Significant lowering in our defect rate?

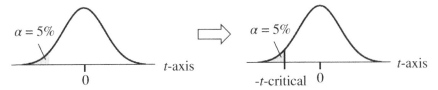

<u>2-tailed test:</u> Has there been a Statistically Significant change (in either direction) from the historical Process Mean?

For a 2-tailed test, Alpha is split in half, and the halves are positioned under the left and right tails of the curve.

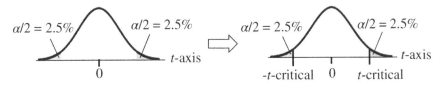

The *t*-Distribution is symmetrical about its Mean of zero. *t*-critical is the distance from the boundary of a shaded Alpha area to zero. For a 2-tailed test, the half-Alpha areas are smaller than the full Alpha area. So, they **will begin far out from the center**. So, the boundaries marked by *t*-critical and its negative will be farther from the center. That means that **the value of *t*-critical for a 2-tailed Distribution will be larger than that for a 1-tailed.**

Note that **the data values in the Sample have nothing to do with determining the values of Alpha or *t*-critical.** Alpha is a Cumulative Probability which we choose, and *t*-critical is calculated from Alpha and the *t*-Distribution. (We do use the Sample Size, *n*, in identifying which *t*-Distribution to use, but we don't use any data values.)

🔑 | **5. Or use the appropriate *t*-Distribution and *t* to determine *p*.**

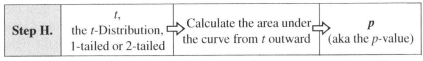

| **Step H.** | *t*, the *t*-Distribution, 1-tailed or 2-tailed | ⇨ Calculate the area under the curve from *t* outward | ⇨ *p* (aka the *p*-value) |

If we're going to make an yes or no conclusion – that is, the difference is or is not Statistically Significant – then Step G and Step H are redundant. Comparing *t* to *t*-critical gives us exactly the same answer as comparing *p* to Alpha.

However, as explained in the article *"p, p-value"* some testers use the *p*-value to define different gradations of how strong the evidence is against the Null Hypothesis. For example, $p < 0.05$ may be considered "moderate" evidence against, while $p < 0.01$ is considered "strong" evidence.

The situation with p and t is kind of the reverse of the situation with Alpha and t-critical. **To find the value of p, we start with the value of t**, which forms the boundary for an area under the curve of the appropriate t-Distribution. p, **then, is the Cumulative Probability of the area under the curve along a tail which is bounded by t.**

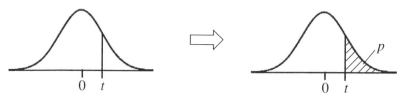

p **is the Probability of an Alpha Error.** (An Alpha Error is also known as a Type I Error, or a **False Positive**. It is the Probability that the test will falsely conclude that there is a Statistically Significant difference between the two Means, when in fact there is not a Statistically Significant difference.)

Exactly like Alpha and t-critical, **t and p convey the same information.** If you know the value of one and which t-Distribution to use, you can calculate the value of the other.

> **6. If $p \leq \alpha$ (equivalently, if $t \geq t$-critical), then there is a Statistically Significant difference between the Means. Otherwise, there is not.**

Step I.	t and t-critical $\Rightarrow$ α and p	Is $t \geq t$-critical? or (same thing) Is $p \leq \alpha$? $\Rightarrow$	Yes: There <u>is</u> a difference; reject the Null Hypothesis (H_0) No: There is <u>no</u> difference; Accept/Fail to Reject H_0

The shaded area representing Alpha is sometimes called the Rejection Region. If t is in this region – as in the left graph below – then the Null Hypothesis is Rejected. Conversely, the unshaded area outside Alpha $(1 - \alpha)$ is called the Acceptance Region. If t is in this area – as in the right graph – we Accept (Fail to Reject) the Null Hypothesis.

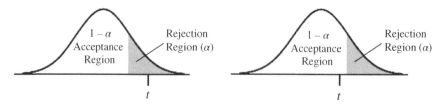

If $p \leq \alpha$ or $t \geq t$-critical   (these two comparisons are statistically identical)

We said earlier that "*t* is a measure of how likely it is that a difference in Means is Statistically Significant." So the larger the value of *t*, the more likely it is that the difference is Statistically Significant (and that we reject the Null Hypothesis).

What factors make *t* larger? Let's look at the generic formula for *t*:

$$t = \frac{\textbf{difference between two Means}}{\textbf{Standard Error}}$$

So, a larger difference between the two Means and/or a smaller Standard Error would make *t* larger. A generic formula for Standard Error is:

$$\textbf{Standard Error} = s/\sqrt{n}$$

This gives us:

$$t = \frac{(\text{difference between two Means})(\text{square root of Sample Size})}{\text{Standard Deviation}}$$

So, **these things will make *t* larger:**
- **a larger difference in the Means**
- **a larger Sample Size**
- **less Variation in the Sample data**

And if *t* **is large enough, ($t \geq t$-critical), then we conclude that there is a Statistically Significant difference, that is, we Reject the Null Hypothesis.**

(Alpha is the shaded area in the diagrams below; *p* is the hatched area.)

Means <u>are</u> different

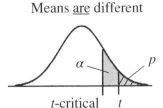

A larger *t* means that *t* is farther away from the Mean and farther out on the tail. This makes *p*, the area bounded by *t* smaller. You can see how *t* > *t*-critical means that *p* (the hatched area) must be less than Alpha (shaded area). (Note, to avoid the visual becoming too cramped, the Alpha pictured above is considerably larger than the usual 5%)

If $p > \alpha$ or $t < t$-critical (these two comparisons are statistically identical)

Means are <u>not</u> different

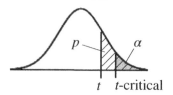

t t-critical

If the calculated value of t is $< t$-**critical**, then t is inside the Acceptance Region – the unshaded area which includes the center of the Distribution. t being smaller than t-critical makes p larger than Alpha.

We conclude that **there <u>is not</u> a Statistically Significant difference** between the two Means we are comparing. In Hypothesis Testing, we **Accept (Fail to Reject) the Null Hypothesis** (which states there is no difference).

Confidence Intervals

Hypothesis Testing and Confidence Intervals are the two major methods of Inferential Statistics. Instead of comparing p to Alpha, or t to t-critical, as in Hypothesis Testing, **it is logically and statistically equivalent to use Confidence Intervals.** These are provided by the software used in t-tests. Confidence Intervals are in units of the data, x, not t. (See the articles, *Confidence Intervals – Parts 1* and *2*.)

t-test	If …		then
1-Sample	the specified Mean	is within the Confidence Interval	there is <u>not</u> a Statistically Significant difference.
2-Sample	zero		
Paired			

Related Articles in This book: *t-tests – Part 1: Overview*; *t, the Test Statistic and Its Distributions*; *Alpha, p-Value, Critical Value and Test Statistic – How They Work Together*; *Confidence Intervals – Parts 1 and 2*; *Critical Values*; *Degrees of Freedom*; *Hypothesis Testing*; *Null Hypothesis*; *Standard Error*; *Standard Deviation*; *p, t, and F: ">" or "<" ?*; *Alpha and Beta Errors*

TEST STATISTIC

Summary of Keys to Understanding

 1. A **Statistic** is a numerical property calculated from Sample Data. A **Test Statistic** is one which has an associated **Probability Distribution**.

 2. There are four commonly used **Test Statistics**: z, t, F, and χ^2 (**Chi-Square**). They are used in a variety of tests in **Inferential Statistics**.

Test Statistic	Used for
z	Comparing Proportions, Comparing Means
t	Comparing Means
F	Comparing Variances
χ^2	Comparing Variances, Determining Independence, Determining Goodness of Fit

 3. A <u>higher value</u> for the Test Statistic tells us that the Sample is likely to be <u>more accurate</u> as a representative of the Population or Process as a whole.

 4. The calculated value for a Test Statistic is a point on the horizontal axis of the Test Statistic's Distribution. It marks the boundary for p, the Probability of an Alpha Error.

 5. If **Test Statistic** ≥ **Critical Value** (this is statistically identical to $p \leq \alpha$), **we conclude that there is a Statistically Significant difference, change, or effect.** That is, we Reject the Null Hypothesis.

Explanation

This article is an overview of the concept of Test Statistic. There are individual articles providing additional detail on each of the four Test Statistics covered here.

 1. **A Statistic is a numerical property calculated from Sample Data. A Test Statistic is one which has an associated Probability Distribution.**

A Statistic distills the information contained in the multiple data values of a Sample into a single number which is used to describe the Sample. For example, the Mean is a Statistic which describes the center of the Sample's data, and the Standard Deviation describes how widely spread the data are. But, if we know the value of one of these Statistics in a Sample, say the Mean = 172 cm, that's the end of the story.

A Test Statistic, on the other hand, is associated with a Probability Distribution or a family of such Distributions. Given a value for the Test Statistic, we can use the Distribution to tell us how likely that value is. It gives us a Probability. It also gives us a Cumulative Probability for all the values less than or greater than that value of the Test Statistic. This is very useful in Inferential Statistics, as we will see.

 2. **There are four commonly used Test Statistics: z, t, F, and χ^2 (Chi-Square). They are used in a variety of tests in Inferential Statistics.**

Test Statistic	Used for
z	**Comparing Proportions, Comparing Means**
t	**Comparing Means**
F	**Comparing Variances**
χ^2	**Comparing Variances, Determining Independence, Determining Goodness of Fit**

In Inferential Statistics, unlike Descriptive Statistics, we don't have access to all the data in the Population or Process. So, we take a Sample and use that to estimate a property of the Population or Process. Being an estimate, there is a Probability of error. **The Probability Distributions associated with Test Statistics enable us to precisely determine the**

Probability of error inherent in our estimates from Sample data. This is why Test Statistics are so valuable, they bring known Probabilities to the analysis.

z is the Test Statistic associated with the Standard Normal Distribution,

the Normal Distribution with a Mean of 0 and a Standard Deviation of 1. **If we assume that the *x* data in the Population or Process follow this Distribution, we can use *z* to tell us what is the Probability of exceeding (or being less than) a given value for *x*.**

And, as described in the article *Proportions*, *z* **can also be used to solve problems involving Proportions of 2-category Count data.**

z can also be used in determining whether there is a Statistically Significant difference between a Sample Mean and a specified Mean. However, problems involving Means are better solved using the *t* Test Statistic. This is true when the Population or Process Standard Deviation (σ) is not known and for "small" Sample Sizes, e.g., $n < 30$ (some experts say when $n < 100$).

z **is unique in only having one associated Distribution. The other Test Statistics are all associated with families of Distributions**, a different Distribution for each different value of Degrees of Freedom (which is calculated from the Sample Size).

t is the preferred Test Statistic for solving problems involving Means.

The formula for *t* involves the Sample Size, *n*. This means that there are different *t*-Distributions for different values of *n*. Consequently, *t* self-adjusts for small Sample Sizes. This is why *t* should be used instead of *z* for problems involving Means when the Sample Size is small ($n < 30$, although some say $n < 100$).

There are three different *t*-tests, differentiated by the types of Means that they compare:

t-test	Means being compared
1- Sample	Sample Mean to a specified Mean
2-Sample	Means of Samples from two different Populations or Processes
Paired	Mean of the differences in pairs of measurements to a Mean of zero

F is the Test Statistic for comparing two Variances.

F is simply the ratio of the Variances of two Samples. **The *F*-test will tell us whether there is a Statistically Significant difference between**

the two Variances. This is analogous to what the 2-Sample *t*-test does with Means.

Data sets are often succinctly described by giving their Mean (for a description of the center) and Standard Deviation (to describe the Variation). Variance is the square of the Standard Deviation. So, often a *t*-test and an *F*-test can be used to determine if there is a Statistically Significant difference between the two sets of data.

"Equal" Variance is also a prerequisite for using a number of statistical tests. The *F*-test can be used to determine this.

Also, the *F*-test is used – in a very creative way – to determine if there is a Statistically Significant difference among three or more Means. See the article *ANOVA – Part 2: How it Does It*.

Chi-Square is a very versatile Test Statistic used in three tests.

The Chi-Square Test for the Variance is analogous to the 1-Sample *t*-test. It compares the Variance calculated from a Sample with a specified Variance. The Variance we specify could be a target value, a historical value, or an estimate.

The Chi-Square Test for Goodness of Fit can be used to determine whether Sample data
  – fit a specified set of values (e.g., our predicted values)
  – fit a Discrete or Continuous Distribution
The Chi-Square Test for Independence can be used to determine whether two Categorical Variables are Independent or Associated. (e.g., gender and fruit juice preference.)

There are separate articles in this book for each of these three tests.

> **3. A higher value for the Test Statistic tells us that the Sample is likely to be more accurate as a representative of the Population or Process as a whole.**

Here **we're defining "more accurate" as** having a lower Probability of an Alpha (False Positive) Error. That is, it has a **lower Probability of leading to a conclusion that there is a Statistically Significant Difference, when – in reality – there is not.**

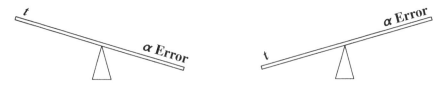

To illustrate, let's look at the formula for the Test Statistic t in the 2-Sample t-test

$$t = \frac{(\textbf{difference between two Sample Means})\sqrt{n}}{s}$$

Let's start with the difference between the two Means in the numerator. It stands to reason that, it would be highly unlikely to have a huge difference between the data values in the two Samples if there were no difference in the two Populations.

The Sample Size is also in the numerator (although the magnitude of its influence is mitigated by the square root). It also makes sense that a larger Sample would be less likely to be very unrepresentative of the Population.

The Standard Deviation, s, is in the denominator. So, the smaller the Variation in the data, the more consistent it is, and thus the more likely that the Sample would be representative.

All this is an attempt to explain in common-sense fashion why a Test Statistic is a measure of how accurate our Sample is likely to be. Let's now look at the statistics of it.

> **4. The calculated value for a Test Statistic is a point on the horizontal axis of the Test Statistic's Distribution. It marks the boundary for p, the Probability of an Alpha Error.**

Using a formula like the one for t shown above, we **calculate the value of the Test Statistic from the Sample data**. We plot this value on the horizontal axis of the Test Statistic's Distribution. **This marks the boundary for the area under the curve beyond** (farther from the Mean than) **the calculated value of the Test Statistic. This area is p, the Probability of an Alpha** (False Positive) **Error.**

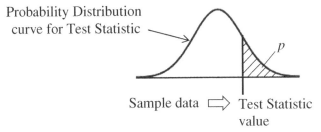

Probability Distribution curve for Test Statistic

p

Sample data ⇨ Test Statistic value

You can see that, if the Test Statistic is larger, the boundary it forms is farther to the right, resulting in the area representing p being smaller. This is how a larger value for the Test Statistics results in a

smaller Probability of Alpha Error. Conversely, a smaller value for the Test Statistic moves the boundary to the left, resulting in a larger p, a larger Probability of Alpha Error.

> **5. If Test Statistic ≥ Critical Value** (this is statistically identical to $p \leq \alpha$), **we conclude that there is a Statistically Significant difference, change, or effect,** that is, we Reject the Null Hypothesis.

How large does the Test Statistic need to be in order for us to conclude that any difference, change, or effect observed in the Sample is Statistically Significant for the Population or Process as a whole? Larger than its Critical Value.

In Inferential Statistics, one of the first steps is to select a value for Alpha (α), the Level of Significance. Most commonly, 5% is selected. This is a clip level for p. It defines the boundary between those values of p which lead to a conclusion of Statistically Significant and those that do not. So, as shown in the table below, we could compare p to α to come to a conclusion.

Alternately, we could take the value for Alpha, plot it as an area under the curve, and calculate its boundary. This boundary is the Critical Value (t-critical below).

Comparing the Cumulative Probabilities (areas under the curve) α and p is statistically identical to comparing the point values of the Test Statistic and the Critical Value of the Test Statistic. The table and diagrams below illustrate all this.

Areas under the curve (right tail) α:▭ p:▨	t-critical t	t t-critical
	$p \leq \alpha$ $t \geq t$-critical	$p > \alpha$ $t < t$-critical
The observation from the Sample data is an accurate estimate for the Population or Process as a whole.	True	False
Null Hypothesis	Reject	Accept (Fail to Reject)
The observed difference, change, or effect is:	Statistically Significant	not Statistically Significant

To reinforce your understanding of how these concepts work together, it may be a good idea to read the article *Alpha, p, Critical Value, and Test Statistic – How They Work Together*.

Related Articles in This Book: *Inferential Statistics*; *Distributions – Parts 1–3*; *z*; *Normal Distribution*; *Proportion*; *F*; *ANOVA – Part 2*; *t – the Test Statistic and its Distributions*; *t-tests – Parts 1 and 2*; *Chi-Square—the Test Statistic and its Distributions*; *Chi-Square Test for Goodness of Fit*; *Chi-Square Test for Independence*; *Chi-Square Test for the Variance*; *Alpha, α*; *p, p-Value*; *Critical Value*; *Alpha, p, Critical Value, and Test Statistic – How They Work Together*; *p, t, and F: "<" or ">" ?*

VARIABLES

Summary of Keys to Understanding

 1. **A Variable is a fundamental mathematical construct which represents entities that can be measured or counted, resulting in more than one value of the Variable.**

 2. **Continuous Variables (aka "Measurement" or "Variables" Variables) represent entities that can be measured, resulting in Continuous data.**

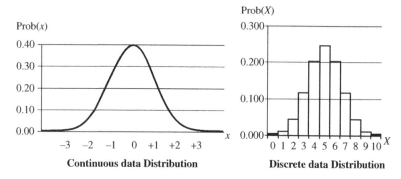

Continuous data Distribution	Discrete data Distribution

 3. **Categorical Variables (aka "Nominal" or "Attributes" Variables) represent entities that can be counted, resulting in Discrete data.**

Categorical Variable	gender	
values of the Variable (the categories)	female	male
Counts within each category	53	47

 4. **A Dependent (aka Response) Variable, y, is a function of one or more Independent (Factor) Variables, x's.**

$$y = f(x_1, x_2, \dots, x_n)$$

5. **A Random Variable is one whose values are determined by chance.**

Statistics from A to Z: Confusing Concepts Clarified, First Edition. Andrew A. Jawlik.
© 2016 John Wiley & Sons, Inc. Published 2016 by John Wiley & Sons, Inc.

Explanation

> **1. A Variable is a fundamental mathematical construct which represents entities that can be measured or counted, resulting in more than one value of the Variable.**

If you look up "Variable" in five different books or websites, you may find five different definitions saying somewhat different things. Since there does not appear to be an official or consensus definition, let's try the one above.

Given how ill-defined and widely encompassing this concept is, our definition has to be somewhat abstract. But don't worry, we get down into the details and examples soon enough.

A Variable is a fundamental ...

"Fundamental" is used to distinguish a Variable from a function, which might otherwise also fit the definition. (In this usage, a function, like $y = a_1 x_1 + a_2 x_2 + b$, is not fundamental, because it can contain Variables.)

... mathematical construct which represents entities ...

A Variable does not exist in the real world. It can be used to represent things that do. It can also be used to represent entities that don't, such as other mathematical constructs.

... that can be measured or counted,

Numerical Variables, like length or weight, can be measured. But even non-numerical Categorical Variables, like gender (which has the non-numerical values of "female" and "male"), can give us numerical Counts.

... resulting in more than one value of the Variable.

That's the whole point of a Variable – it represents entities that can vary in value. Otherwise it would be a Constant, not a Variable.

> **2. Continuous Variables** (aka "Measurement" or "Variables" Variables) **represent entities that can be <u>measured</u>, resulting in Continuous data.**

Examples are length, weight, and temperature. Theoretically, Continuous Variables can take on an infinite number of values between any two values. This would give a smooth curve for a Probability Distribution. Practically speaking, things that have a lot of values without being infinitely divisible (e.g., money) can be treated as Continuous Variables/data.

The main reason that there is a distinction between Continuous and Discrete data is that two different types of Probability Distributions are used for the two different types of Variables/data. See the article, *Distributions – Part 3: Which to Use When.*

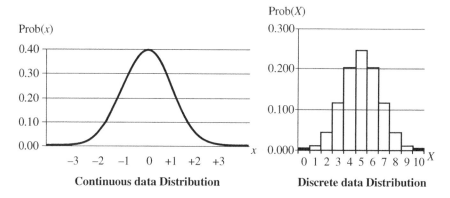

Continuous data Distribution   Discrete data Distribution

 3. Categorical Variables (aka "Nominal" or "Attributes" Variables) **represent entities that can be <u>counted</u>, resulting in Discrete data.**

Let's say we're interested in determining whether there was a Statistically Significant difference between the number of men and women who shopped at a certain store. We took a Sample of 100 people as they entered the store.

Categorical Variable	gender	
values of the Variable (the categories)	female	male
Counts within each category	53	47

- The Categorical <u>Variable</u> is gender.
  It classifies the test subjects by a non-numerical Attribute.
- The <u>values</u> of the Variable are the names (hence "Nominal") of the categories.
- The <u>data</u> is the Count within each category.

Count data can be collected for Categorical Variables. Count data are Discrete; they can only consist of non-negative integers, such as 0, 1, 2, 3 …

Ordinal Variables are a type of Categorical Variable. In this type, the category names convey an implicit ordering, but there is no explicit numerical value associated with each category name. Examples are "good," better," "best" or "beginner," "intermediate," "advanced."

Numerical and non-numerical: Variables are sometimes classified into two groups – numerical and non-numerical. Categorical Variables are classified as non-numerical, because their values are category names, not numbers. However, this can be confusing, because, as we have seen, numerical data, in the form of Counts, are collected for Categorical Variables.

Bins can be used to change Continuous data into Counts of a Categorical Variable.

The Bins can be names of ranges, e.g., Continuous data: weight in pounds. The values of the Categorical Variable called "Weight" would then be the names of the ranges:

categories (bins)	"<100"	"101 to 125"	"126 to 150"	"151 to 200"	">200"
Counts	22	49	117	53	19

> **4. A Dependent (aka Response) Variable, y, is a function of one or more Independent (Factor) Variables, x's.**
>
> $$y = f(x_1, x_2, \dots, x_n)$$

We run into this classification of Variables in ANOVA, Regression, and Design of Experiments. There is a cause (Independent Variables) and an effect (Dependent Variable) relationship implied. This is a stronger relationship than Correlation or Association.

The Independent Variable(s) (the x's) can take on values independently. But the value of the y Variable is Dependent on the values of the x Variable(s).

Synonyms:

x	Independent Variable	Cause	Predictor Variable		Explanatory Variable
y	Dependent Variable	Effect	Outcome Variable	Response Variable	Criterion Variable

> **5. A Random Variable is one whose values are determined by chance.**

For example, the number of heads in 10 coin flips or the number of shoppers entering a store in a 15-minute interval. Random Variables can have either Discrete or Continuous data. So all data collected in random Samples could be expressed as Random Variables.

A Binomial Random Variable is one that has only two values, for example, heads or tails, green or not green, defective or not defective, and which meets the other requirements for using the Binomial Distribution. See the article *Binomial Distribution*.

Related Articles in This Book: *Distributions – Part 1: What They Are*; *Distributions – Part 3: Which to Use When*; *Binomial Distribution*; *Chi-Square Test for Independence*; *Regression – Part 2: Simple Linear*

VARIANCE

Symbols: σ^2, s^2, $V(x)$, $VAR(x)$, $Var(x)$

Summary of Keys to Understanding

 1. **Variance is a measure of Variation about the Mean.**

 2. **Variance of a Population or Process:** $\quad \sigma^2 = \dfrac{\sum (x_i - \mu)^2}{N}$

Variance of a Sample: $\quad s^2 = \dfrac{\sum (x_i - \bar{x})^2}{n - 1}$

There are also formulas for the Variance of Random Variables.

 3. **Variance is of limited practical use, because it is expressed in units which are squares of the units of the data,** e.g., square kilograms, square gallons, square IQ points.

   The Standard Deviation is the square root of the Variance.

   It is more useful and more used than Variance, **because it is in the same units as the data,** e.g., kilograms, gallons, IQ points.

 4. **The Squaring is done to convert negative differences** (for values less than the Mean) **into positive numbers,** so that they don't cancel out the positive differences.

   Absolute Values would do the same thing, but there are advantages to squaring.

 5. **The F-test compares the Variances from two Samples.**

   The Chi-Square Test for the Variance compares the Variance calculated from one Sample of data to a Specified Variance.

6. **Equal Variance** – which can be determined via the F-Test or Levene's Test – **is a requirement for being able to perform a number of Inferential Statistical analyses.**

Explanation

 | **1. Variance is a measure of Variation about the Mean.** |

Variation (also known as Variability, Dispersion, or Spread) is one of three categories of descriptors for a Population or a Distribution or a Sample. The other two categories are Central Tendency (e.g., Mean, Mode, Median) and Shape (Skewness and Kurtosis).

Variance is one of several measures of Variation. Others include Range and Standard Deviation (the latter of which is derived from Variance).

Let's look at three admittedly unusual Populations in order to demonstrate a few points.

Population A: $-50, -40, -30, -20, -10, 0, 10, 20, 30, 40, 50$

Population B: (5 values at -50 and 5 at 50)

Population C: $-50, -20, -20, -10, -10, -10, 0, 0, 0, 0, 10, 10, 10, 20, 20, 50$

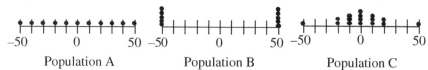

| Population A | Population B | Population C |

The first thing we might notice is that, while they are very different, they have the same Mean: 0. So, a measure of Central Tendency like the Mean is not sufficient to describe a Distribution. Adding a measure of Variation (also known as Variability, Dispersion, or Spread) would be helpful. But which measure of Variation?

Range is a somewhat limited measure. It defines the boundaries of a set of values, but it doesn't tell you much about what happens in between. For example: Populations A, B, and C all have the same Range: 100 (from -50 to $+50$). But they are very different in whatever other definition of Variation we may want to have.

We need a more descriptive measure of Variation. **It would be good if we could somehow quantify the degree of "clustering" around the Mean.** Clustering is good to emphasize, because it measures how accurate a Sample is likely to be in estimating a property of a Population. (And this is the central purpose of Inferential Statistics.)

How might we go about this? Maybe we could calculate the average distance of the data values from the Mean.

Average distance From the Mean $= (\Sigma(x_i) - \mu)/N$

Since the Mean (μ) is zero for Populations A, B, and C, we would be subtracting zero. So, we just sum the individual data values. Here we'll re-sequence the individual values to make a point more obvious:

Sum of the data Values:

Population A: $-50 + 50 + -40 + 40 + -30 + 30 + -20 + 20 + -10 + 10 + 0 = 0$

Population B: $-50 + 50 + -50 + 50 + -50 + 50 + -50 + 50 + -50 + 50 = 0$

Population C: $-50 + 50 + -20 + 20 + -20 + 20 + -10 + 10 + -10 + 10 + -10 + 10 + 0 + 0 + 0 + 0 = 0$

No need to go any farther; we can see this is not going to work. The average distance from the Mean will always be zero. In retrospect, this is not surprising, since the very nature of the Mean guarantees that this will happen. Some values will be less than the average (yielding negative values for differences) and some more (yielding positive values). And these negatives and positives average each other out.

There are two ways to change negatives to positives – square them or use their absolute value. The Variance uses the squaring method. The reason why is explained in Key to Understanding #4.

> **2. Variance of a Population or Process:** $\qquad \sigma^2 = \dfrac{\sum (x_i - \mu)^2}{N}$
>
> **Variance of a Sample:** $\qquad s^2 = \dfrac{\sum (x_i - \bar{x})^2}{n - 1}$
>
> There are also formulas for the Variance of Random Variables.

Where

σ is the Standard Deviation of the Population or Process

s is the Standard Deviation of the Sample

x_i represents the individual data values

μ (mu) is the Population or Process Mean

$\bar{x}$ is the Sample Mean

N is the number of items in the Population or Process

n is the number of items in (the Size of) the Sample

Note that the symbols for the Variance are the squares of the symbols for Standard Deviation.

The formula for the Population or Process divides by N. But the formula for the Sample divides by the Sample Size minus 1. This gives us a somewhat larger value for any given Sample Size. How much larger is dependent on the Sample Size. The effect on the Variance of using $n - 1$ is larger for small Sample Sizes.

n	$n - 1$	% difference
5	4	20%
100	99	1%

The $n - 1$ adjustment accounts for the fact that small Samples will give less accurate estimations of the Population or Process Standard Deviation than will larger Samples. So, **Variances of small Samples will generally be larger than those of large Samples.**

There are also formulas for the Variance of Random Variables. The formulas above are for data values which we can measure in a Population, Process, or Sample. Random Variables take on values due to chance. They are described by Probability functions. So, the formulas for Variance can use these Probability functions, $P(x)$, to predict values for summing.

Variance of a Discrete (countable) Random Variable:

$$\mathbf{Var}(X) = P(X_i)\Sigma(X_i - \mu)^2$$

where μ is the Expected Value.

Variance of a Continuous Random Variable:

Continuous variables can take on an infinite number of values, so they can't be summed like Discrete Variables. The <u>integral</u> is used in place of the sum:

$$\mathbf{Var}(x) = P(x_i) \int (x_i - \mu)^2$$

Variance of a Binomial Random Variable:

$$\mathbf{Var}(x) = \mathbf{np(1 - p)}, \text{ where } p \text{ is the Proportion}$$

> **3. Variance is of limited practical use, because it is expressed in units which are squares of the units of the data**, e.g., square kilograms, square gallons, square IQ points.
>
> **The Standard Deviation is the square root of the Variance.**
>
> **It is more useful** and more used than Variance, **because it is in the same units as the data**, e.g., kilograms, gallons, IQ points

These squared units have either no meaning – e.g., square kilograms, square gallons, square IQ points, or they are misleading – e.g., square kilometers. As a result, Variance often serves only as an interim point in the calculation of the Standard Deviation, which is the square root of the Variance. The Standard Deviation is in the same units as the data – e.g., kilograms, gallons, IQ points, or kilometers.

However, special types of Variances appear as interim steps in calculations in some statistical analyses. For example, if we use a generic description of Variance as

Variance = Sum of Squares/Degrees of Freedom,

then Mean Sums of Squares Within (MSW) and Mean Sums of Squares Between (MSB) are two types of Variances used in ANOVA. See the article *ANOVA – Part 2: How it Does It.*

---

4. The Squaring is done to convert negative differences (for values less than the Mean) **into positive numbers,** so that they don't cancel out the positive differences.

Absolute Values would do the same thing, but there are advantages to Squaring.

Why not just use the absolute values instead of squaring? After all, one usually reverses the squaring in the Variance by taking the square root to calculate the Standard Deviation. It makes sense to think of the absolute value first, when you want to change negative numbers to positive numbers. In fact, there are measures – variously called the Average Absolute Deviation or the Mean Absolute Deviation – which do just that. Compared to Variance, these downplay the effect of outliers. And that may be desirable for some uses.

By using squares, Variance confers an outsized significance on values far from the Mean. In order to have a relatively low value for Variance, the values have to be clustered close to the Mean. Data points far outside have a disproportion impact on the value of the Variance. In addition to avoiding square units, this is another reason why Standard Deviation is used more often than Variance. The Variance's overemphasis of far-out values can be mitigated to a significant degree by taking the square root of the Variance, yielding the Standard Deviation.

The following table illustrates this with some admittedly extreme Populations. Note that Populations D and E have exactly the same data values, with one exception: Population D has one outlier at −50. This makes its Range triple that of Population E, and its Variance is almost four times as much. But its Standard Deviation is only about twice as much.

	Population A $\underset{-50 \quad\quad 0 \quad\quad 50}{\text{├┼┼┼┼┼┼┼┼┼┼┤}}$	Population D $\underset{-50 \quad\quad 0 \quad\quad 50}{\text{├┼┼┼┼┼┼┼┤}}$	Population E $\underset{-50 \quad\quad 0 \quad\quad 50}{\text{├┼┼┼┼┼┼┼┤}}$
Range	100	60	20
Avg. Absolute Dev.	27.3	8.7	5.7
Variance	1000	220	57
Standard Deviation	32	15	7.6

But, **the main reason for using squaring is that, it is simpler to mathematically manipulate squares than absolute values**. This is due to the so-called "Pythagorean" Theorem of Statistics. For Variances, this says that, if X and Y are independent Random Variables, then

$$\text{Var}(X \pm Y) = \text{Var}(X) + \text{Var}(Y)$$

So, a number of statistical formulas and calculations use the Variance through the initial calculation steps and then calculate the Standard Deviation from the Variance.

> **5. The F-test compares Variances from two Samples.**
>
> **The Chi-Square Test for the Variance compares the Variance calculated from one Sample of data to a Specified Variance.**

If we want to know whether there is a Statistically Significant difference in the Variation of two Populations or Processes, we would use the F-test. The F-Test Statistic is simply the ratio of the two Variances.

If, however, we're interested in the Variation in one Population or Process, and we want to know whether there is Statistically Significant difference between that Variance and a specified value, then we would use the Chi-Square Test for the Variance. The specified value could be something like a target – say, a national average (for which we would not have the underlying data) – or a historical level, say the usual Variation in a manufacturing process.

If you're familiar with t-tests, it may help to think of the F-test as being analogous to the 1-Sample t-test, while the Chi-Square Test for the Variance is analogous to the 2-Sample (aka Independent Samples) t-test.

6. Equal Variance – which can be determined via an *F*-Test or Levene's Test – **is a requirement for being able to perform a number of Inferential Statistical analyses.**

… including ANOVA, the 2-Sample *t*-test, Residuals in Multiple Regression.

Related Articles in This Book: *Variation/Variability/Dispersion/Spread*; *Standard Deviation*; *Binomial Distribution*; *ANOVA – Part 2: How it Does It*; *F*; *Chi-Square Test for the Variance*; *Nonparametric*

VARIATION/VARIABILITY/ DISPERSION/SPREAD

Summary of Keys to Understanding

 1. Variation (also known as Variability, Dispersion, and Spread) **is one of three major** <u>categories of measures</u> **describing a Distribution or data set.**

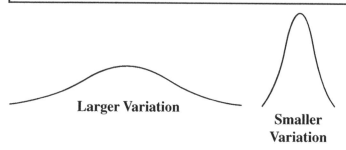

Larger Variation

Smaller Variation

 2. There are a number of different measures of Variation – each with its pros and cons.

	Range	InterQuartile Range (IQR)	Variance	Mean Abs. Deviation (MAD)	Standard Deviation
Effect of very high or very low values	Is defined by highest and lowest	None	Overly emphasized by Squaring	Handled the same as other values	Somewhat dispropor- tionate
Identifies Clustering around Mean	N	Y	Y	Y	Y
Is in Units of the data	Y	Y	N	Y	Y
Use	Least useful in statistics	In Box-and- Whiskers Plot	For calculating Standard Deviation	Least Common	Most Common

 3. Distributions are often succinctly described by stating the Mean (for Central Tendency) **and the Standard Deviation** (for Variation).

Explanation

> **1. Variation** (also known as Variability, Dispersion, and Spread) **is one of three major <u>categories of measures</u> describing a Distribution or data set.**

A fifth synonym is "Scatter." The other two categories are Central Tendency (Mean, Mode, Median) and Shape (Skew and Kurtosis).

The Distribution can be of a Population, Process, Sample, or other data set.

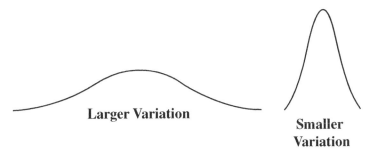

Larger Variation

Smaller Variation

> **2. There are a number of different measures of Variation – each with its pros and cons.**

	Range	**InterQuartile Range (IQR)**	**Variance**	**Mean Abs. Deviation (MAD)**	**Standard Deviation**
Effect of very high or very low values	Is defined by highest and lowest	None	Overly emphasized by Squaring	Handled the same as other values	Somewhat disproportionate
Identifies Clustering around Mean	N	Y	Y	Y	Y
Is in Units of the data	Y	Y	N	Y	Y
Use	Least useful in statistics	In Box-and-Whiskers Plot	For calculating Standard Deviation	Least Common	Most Common

This is not an exhaustive list.

Range:

Range is simply the difference between the highest and lowest values.

It may be the least useful in statistics. It only tells you about two values – out of the many which may be in the Distribution or dataset. It tells you nothing of the values in between the highest and lowest values.

InterQuartile Range (IQR):

The InterQuartile Range provides information on 50% of the data values, which is why it is also called the **"middle 50."** It is the Range of the values around the Mean which comprise 50% of the total values.

The lower boundary value of the IQR box in the diagram below is called the **25th percentile,** and the upper boundary value is called the **75th percentile.** This is because 25% (one quarter) of the Distribution's values are below the lower limit of the IQR, and 25% are above the upper limit of the IQR. **The 50th percentile** is, by definition, the Median.

The IQR is used to define Outliers and Extremes.

IQRs are often depicted via Boxplots – or Box-and-Whiskers Plots, such as the one below.

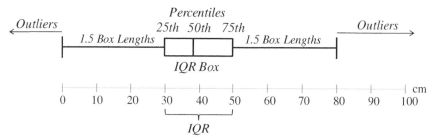

The box defines the boundaries of the "middle fifty." The IQR (Box Length) in this example is 20 cm (50 – 30). The thickness (height) of the box is meaningless; it just serves to make the rectangular shape that differentiates the box from the "whiskers" to the left and right.

1.5 box length is 30 cm. **Outliers** and **Extremes** are any values outside 1.5 box lengths and 3 box lengths, respectively, beyond from the 25th and 75th percentile.

The Box-and-Whiskers Plot is very useful for conveying a lot of information visually. Showing several vertically oriented Boxplots together is a good way to compare the Variations of several data sets. See the article *Charts, Graphs, Plots – Which to Use When.*

Variance:

There is a separate article on Variance. But briefly, it is the average of the squares of the distances of each data value from the Mean. Its units are the square of the data units (e.g., square gallons, square degrees Centigrade, etc). As a result it is not very useful by itself. Its main use is as an interim step in calculating the Standard Deviation – which is its square root.

Mean Absolute Deviation (MAD):

MAD is the average (unsquared) distance of the data points from the Mean. It is useful when it is desirable to avoid emphasizing the effects of outliers. But it is not very common. It is in the same units as the data.

Standard Deviation:

The Standard Deviation is the most commonly used measure of Variation. It is the square root of the Variance. As a result, it is in the same units as the data. See the article *Standard Deviation.*

 3. Distributions are often succinctly described by stating the Mean (for Central Tendency) **and the Standard Deviation** (for Variation).

The Mean is the most common and most useful measure of Central Tendency, and the Standard Deviation is the same for Variation. The two are often quoted together to portray a Distribution.

One reason for this is that the percentages of the values which fall within a given number of Standard Deviations from the Mean have been determined. For a Normal Distribution, they can be stated very precisely. However, for <u>any</u> Distribution, lower-bound estimates are known.

	Percent of Values Found within this Number of Standard Deviations from the Mean			
	1	**2**	**3**	**4**
Normal Distribution (Empirical Rule)	68.5%	95.5%	99.7%	
All Distributions (Chebyshev's Theorem)		>75%	>88.9%	>93.7%

Related Articles in This Book: *Distributions – Part 1: What They Are*; *Variance*; *Standard Deviation*; *Charts, Graphs, Plots – Which to use When*

WHICH STATISTICAL TOOL TO USE TO SOLVE SOME COMMON PROBLEMS

There are similar "Which to Use When" articles for Charts/Graphs/Plots, for Control Charts, and for Distributions. They can be found in this book alphabetically by the topic name.

EXPECTED FREQUENCIES vs. OBSERVED COUNTS	
Problem/Question/Need	**Tool** *(article which describes it)*
Is our prediction of Expected percentages a good fit with the actual Observed data subsequently collected? For example, We predict the following allocation of customers at our bar by day of the week: M – Th 12.5% each:, Fri 30%, Sat 20%	**Chi-Square Test for Goodness of Fit** *(article by the same name)*

FITTING A FUNCTION (line or a curve) to DATA	
Problem/Question/Need	**Tool** *(articles which describe it)*
What is the straight-line ($y = bx + a$) function (Model) that describes the relationship between one independent (Factor) Variable x and the dependent (Response) Variable y? For example, Total crop harvested as a function of acres planted.	First: **Scatterplot** and **Correlation analysis** to verify linear Correlation. *(Charts, Plots, and Graphs – Which to Use When; Correlation – Parts 1 and 2)* Then, **Simple Linear Regression** *(Regression – Part 2: Simple Linear)*
What is the straight-line ($y = b_1x_1 + b_2x_2 + \ldots + b_nx_n$) function (Model) that describes the relationship between multiple independent (Factor) Variables and the dependent (Response) Variable? For example, House price as a function of the number of bedrooms and bathrooms.	First: **Scatterplots** and **Correlation analyses** to verify linear Correlation between each x Variable and the y Variable – and <u>not</u> between x variables *(Charts, Plots, and Graphs – Which to Use When; Correlation – Parts 1 and 2)* Then, **Multiple Linear Regression** *(Regression – Part 4: Multiple Linear)*

Statistics from A to Z: Confusing Concepts Clarified, First Edition. Andrew A. Jawlik.
© 2016 John Wiley & Sons, Inc. Published 2016 by John Wiley & Sons, Inc.

FITTING A FUNCTION (line or a curve) to DATA (*Continued*)	
Problem/Question/Need	**Tool** *(articles which describe it)*
What is the nonlinear function (Model) that fits a curve $y = f(x)$ **to the data?**	**Simple Nonlinear Regression** *(Regression – Part 5: Simple Nonlinear)*
How do I validate a Regression Model? Not with the data used to produce it. A controlled experiment must be used to test predictions from the Model with new data.	**Design of Experiments** *(Design of Experiments, DOE – Parts 1–3)*

INDEPENDENCE of Categorical Variables	
Problem/Question/Need	**Tool** *(articles which describe it)*
Are the Proportions associated with categories of one Categorical Variable influenced by those of a second Categorical Variable? For example, Is the preference for a "flavor of ice cream" (which is the Categorical Variable with values of "chocolate," "strawberry," and "vanilla") influenced by gender (the Categorical Variable with values "male" and "female")?	**Chi-Square Test for Independence** *(article by the same name)*

MEANS – Measurement/Continuous data These all assume data which are roughly Normal. For non-Normal data, use the Median.	
Problem/Question/Need	**Tool** *(article which describes it)*
Is this Mean different from a specified Mean? For example: • Is our school's average test score different from the national average? • Has the Mean of a measurement or a defect rate in the Process has changed from its historical value? • Does the Mean reduction in blood pressure meet or exceed the target for this new treatment?	**1- Sample** *t*-**test** *(t-tests – Parts 1 and 2)*

(Continued)

MEANS – Measurement/Continuous data (*Continued*) These all assume data which are roughly Normal. For non-Normal data, use the Median.	
Problem/Question/Need	**Tool** *(article which describes it)*
Are these two Means different? For example: • Are our high school's test scores different from another school's? • Do these two treatments have different effects?	**2-Sample *t*-test** *(t-tests – Parts 1 and 2)*
Are these two Means different for the same subjects? For example, Do individuals perform better after this new training than before?	**Paired *t*-test** *(t-tests – Parts 1 and 2)*
Is there a difference among several (more than two) **Means, compared with each other?** For example, There are three types of training given to our workers. Do they result in different effects on worker performance?	**ANOVA** *(ANOVA – Parts 3 and 4)*
Which of several Means are different from the Overall Mean? For example, Which of several production facilities does significantly better or worse than the others?	**ANOM** (ANOM)

MEDIANS For Non-Normal data, use Medians instead of Means. *(See the article "Nonparametric")*	
Problem/Question/Need	**Nonparametric Test**
Is this Median different from a specified **Median?**	**Wilcoxon Signed Rank**
Independent Samples: **Are these two Medians different?**	**Mann–Whitney**
Paired Samples: **Are these two Medians different?**	**Wilcoxon Signed Rank**
1 Variable: **Is there a difference among several Medians?**	**Kruskal–Wallis**
2 Variables: **Is there a difference among several Medians?**	**Friedman**

PROPORTION	
Problem/Question/Need	**Tool** *(article which describes it)*
Confidence Interval estimate of a Proportion from Sample data	*z* *(Proportion)*
Is there a difference between the Proportions from 2 Populations or Processes? For example, 0.52 of women and 0.475 of men preferred Candidate A	*z* *(Proportion)*
Is there a difference among the Proportions from 3 or more Populations or Processes?	**Chi-Square Test for Independence** *(article by the same name)*

VARIATION	
Problem/Question/Need	**Tool** *(article which describes it)*
Is this Variance (or Standard Deviation) **different from a Specified Variance** (Standard Deviation)**?** For example, Has the Variation in our Process increased from the historical value?	**Chi-Square Test for the Variance** *(article by the same name)*
Are these two Variances different? For example, Two treatments have the same Mean effect. The tie-breaker would be whether one had a significantly smaller Variance – that is, it was more consistent.	***F*-test** *(article: F)*

Z

Summary of Keys to Understanding

1. z is a Test Statistic whose Probabilities do not vary by Sample Size.

2. z has only one associated Distribution, the Standard Normal Distribution.

3. A value of z tells you – in units of Standard Deviations – how far away from the Mean an individual data value is.

4. For Normal or near-Normal Distributions, z can be used to solve problems like:
   – Given a value x, what is the Probability of exceeding x? or not exceeding x?
   – Conversely, given a Cumulative Probability, what value of x defines its boundary?
z can also be used to calculate a Confidence Interval estimate of a Proportion.

For your convenience, here are the z scores for several common values of Alpha:

α	0.025	0.05	0.1	0.9	0.95	0.975
z	−1.960	−1.645	−1.282	1.282	1.645	1.960

5. For analyzing Means, z has several significant similarities to t and some key differences. Use z only for large Samples ($n > 30$) and only when the Population Standard Deviation (σ) is known. Otherwise, use t.

Statistics from A to Z: Confusing Concepts Clarified, First Edition. Andrew A. Jawlik.
© 2016 John Wiley & Sons, Inc. Published 2016 by John Wiley & Sons, Inc.

Explanation

 1. z is a Test Statistic whose Probabilities do not vary by Sample Size.

A Test Statistic is a Statistic (a property of a Sample) with a known Probability Distribution. That is, for each value of the Test Statistic, we know the Probability of that value occurring.

z has several uses:

- For data distributed roughly like a Normal Distribution, z can give you the Cumulative Probability associated with a given value for x. Conversely, given a Cumulative Probability, z can give you the associated value of x.
- z can also be used in analyzing Means, but t is a better choice for that purpose.
- z can solve problems involving Proportions of 2-category Count data. (See the article *Proportion*.)

The common Test Statistics other than z – e.g. t, F, and Chi-Square – have a different Distribution for each value of Degrees of Freedom (which is related to Sample Size). Unlike other Test Statistics, there is no "n" for Sample Size in the formulas for z. So z does not vary with Degrees of Freedom. This is because …

 2. z has only one associated Distribution, the Standard Normal Distribution.

A Normal Distribution is the familiar **bell-shaped curve.** The Probabilities of many properties of Populations and Processes approximate the Normal Distribution.

The z Distribution
-- the Standard Normal Distribution (Mean = 0, Standard Deviation = 1)

The Standard Normal Distribution is an idealized version – the Normal Distribution with a Mean of zero and a Standard Deviation of 1. This is the Distribution for z.

z is the Variable along the Horizontal axis in the Standard Normal Distribution.

Point Probability

As for any other Distribution, the vertical axis is the **Point Probability**. So, **for any value of** z (any point on the horizontal axis), **its Point Probability (the Probability of it occurring) is the height of the curve above** z. In the diagram above, we can see that the Point Probability of $z = 0$ is about 0.4.

Cumulative Probability

A Cumulative Probability is the total of the Point Probabilities for a range of values. It is represented by the area under the curve of the range. For example, we can see that Cumulative Probability of the range $z > 0$ is 50% – half the area under the curve.

Here are useful Cumulative Probabilities for the Standard Normal Distribution:

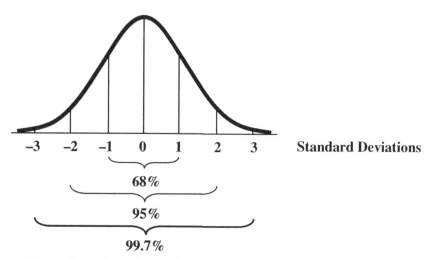

We see that values of z within plus or minus 1 Standard Deviation of the Mean (1σ) occur 68% of the time. Within 2σ, it's 95%; and 99.7% for 3σ.

These percentages were straightforward to calculate for the idealized z-Distribution. But they **can be used for every Normal Distribution**, because of the Empirical Rule:

Empirical Rule (aka the 68, 95, 99.7 Rule): **Given a value – expressed in Standard Deviations – on the horizontal axis of <u>any</u> Normal Distribution, the Probability of that value (vertical axis) is the same for <u>every</u> Normal Distribution**.

Next, we'll see exactly how we use this to find Probabilities for x values in any Normal or near-Normal Distribution.

> **3. A value of z tells you – in units of Standard Deviations, – how far away from the Mean an individual data value, x, is.**

What does an individual value of z tell me?

A value of z is often called a "z-score"

– **It tells you how far away the corresponding data point (x) is from the Mean.**
– **It tells you this in units of Standard Deviations.**

<u>Example</u>: The height of adult males in a Population has a Mean, μ, of 175 cm and a Standard Deviation, σ, of 7 cm.

We use the formula, $z = (x - \mu) / \sigma$ to convert x values into z values (aka z scores).

For $x = 168$ cm, $z = (168 - 175)/7 = -1$; for $x = 175$ cm, $z = (175 - 175)/7 = 0$; for 189 cm, $z = (189 - 175)/7 = +2$

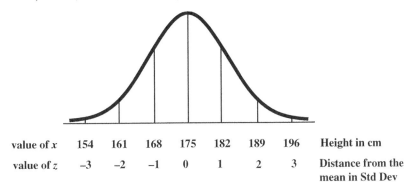

value of x	154	161	168	175	182	189	196	Height in cm
value of z	–3	–2	–1	0	1	2	3	Distance from the mean in Std Dev

> **4. For Normal or near-Normal Distributions, z can be used to solve problems like:**
> – Given a value x, what is the Probability of exceeding x? or not exceeding x?

> – Conversely, given a Cumulative Probability, what value of
> x defines its boundary?
>
> **z can also be used to calculate a Confidence Interval estimate
> of a Proportion.**

Examples of the kind of problems z can be used to solve:

1. The Mean lifetime for a brand of light bulb is 1000 hours, with a Standard Deviation of 100 hours. What <u>percentage</u> (Cumulative Probability) of light bulbs can be expected to burn out <u>before 900 hours (x)</u>?

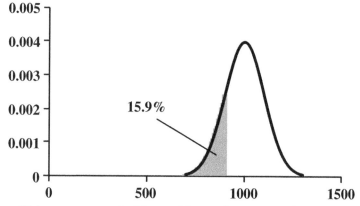

Using z, we can find in a table or via a spreadsheet that the percentage = **15.9%.**

This percentage – of the total area under the curve – is represented by the **shaded area left of $x = 900$.**

2. For the same brand of light bulbs, how many **hours (x)** can we expect **90% (α)** of the light bulbs to exceed?

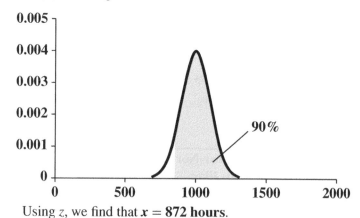

Using z, we find that $x = 872$ **hours.**

The shaded area to the right of 872 hours **comprises 90%** of the area under the curve.

How to do it:
To find a Cumulative Probability for a given *x*, we

1. Convert *x* to its *z* value.
2. Find the Cumulative Probability for that *z* on the Standard Normal Distribution.

To find *x*, given a Cumulative Probability, we

1. Find the *z* value corresponding to that Probability on the Standard Normal Distribution.
2. Convert the *z* to the corresponding *x* value.

It may help to depict these steps graphically: **If you know *x* and want to find its Cumulative Probability:**

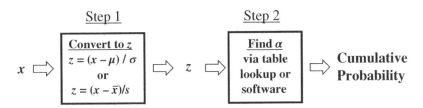

If you know a Cumulative Probability and want to find the corresponding *x*:

z **can also be used to calculate a Confidence Interval estimate of a Proportion.** This is explained in the article *Proportion*.

> **5. For analyzing Means,** *z* **has several significant similarities to** *t* **and some key differences. Use** *z* **only for large Samples (***n* > 30**) and only when the Population Standard Deviation (**σ**) is known. Otherwise, use** *t*.

	z	t
It is a …	Test Statistic	
Probability Distribution(s)	Bell-shaped, Symmetrical, Never touches the horizontal axis	
	1	1 for each value of df
Mean = Mode = Median	Yes, and they all $= 0$	
Formula	$z = (\bar{x} - \mu) / \sigma$	$t = (\bar{x} - \mu) / (s/\sqrt{n})$
Varies with Sample Size	No	Yes
Accuracy	less	more
Standard Deviation	$= 1$	Always > 1

Note that this formula for calculating z from Sample data for use as a Test Statistic is different from the formula for z-score shown earlier.

z is the simplest Test Statistic, and so it is useful to start with it when learning the concept. But it has some significant limitations, due to the fact that it does not take Sample Size into account. There is a much larger risk for errors when a Sample Size of 5 is used vs. 500, but z treats them equally.

z should NOT be used

1. **When the Population is <u>not</u> Normal or near-Normal.**
2. **When the Population Standard Deviation (Sigma, σ) is not known.**
3. **When the Sample Size is small** – some say when $n < 30$, others say when $n < 100$.

As the Sample Size grows larger, the values of z and t converge. For very large Sample Sizes, there is little difference between the values of z and t. As n approaches infinity, the difference between z and t approaches zero. That is why z can be used instead of t for large Sample Sizes.

Related articles in This book: *Test Statistic*; *Normal Distribution*; *Standard Deviation*; *Alpha, α*; *Proportion*; *t – the Test Statistic and its Distributions*

HOW TO FIND CONCEPTS
IN THIS BOOK

This book is alphabetically organized, like a dictionary or an encyclopedia, so an index is not needed.

Readers can quickly find articles on statistical concepts by flipping through the book like a dictionary or a mini-encyclopedia.

The Contents at the beginning of the book lists all the articles on the major concepts.

Immediately following the Contents is a Section called "Other Concepts Covered in the Articles." This lists additional concepts and statistical terms which do not headline an article. For example,

Acceptance Region: See the article *Alpha*, α.

Printed and bound by CPI Group (UK) Ltd, Croydon, CR0 4YY

27/10/2024

14580475-0003